EXPLORING INFORMATION SUPERIORITY

A Methodology for Measuring the Quality of Information and Its Impact on Shared Awareness

Walter Perry

David Signori

John Boon

Prepared for the
Office of the Secretary of Defense

National Defense Research Institute

RAND

The research described in this report was sponsored by the Office of the Secretary of Defense (OSD). The research was conducted in the RAND National Defense Research Institute, a federally funded research and development center supported by the OSD, the Joint Staff, the unified commands, and the defense agencies under Contract DASW01-01-C-0004.

Library of Congress Cataloging-in-Publication Data

Perry, Walt L.
    Exploring information superiority : a methodology for measuring the quality of information and its impact on shared awareness / Walter Perry, David Signori, John Boon.
        p. cm.
    "MR-1467."
    Includes bibliographical references.
    ISBN 0-8330-3489-8 (pbk. : alk. paper)
        1. Command and control systems—United States. 2. Information warfare—United States. I. Signori, David, 1942– II. Boon, John, 1959– III.Title.

UB212.P47 2003
355.3'3041'0973—dc22

                                                            2003022433

The RAND Corporation is a nonprofit research organization providing objective analysis and effective solutions that address the challenges facing the public and private sectors around the world. RAND's publications do not necessarily reflect the opinions of its research clients and sponsors.

RAND® is a registered trademark.

Published 2004 by the RAND Corporation
1700 Main Street, P.O. Box 2138, Santa Monica, CA 90407-2138
1200 South Hayes Street, Arlington, VA 22202-5050
201 North Craig Street, Suite 202, Pittsburgh, PA 15213-1516
RAND URL: http://www.rand.org/
To order RAND documents or to obtain additional information, contact
Distribution Services: Telephone: (310) 451-7002;
Fax: (310) 451-6915; Email: order@rand.org

The military is formulating new visions, strategies, and concepts that capitalize on emerging information-age technologies to provide its warfighters with significantly improved capabilities to meet the national security challenges of the 21st century. These programs are described in such documents as the *Quadrennial Defense Review*, *Joint Vision 2020*, a variety of publications describing network-centric warfare (NCW), and other documents describing military transformation. *Joint Vision 2020* provides an important starting point for describing a future warfighting concept that has since evolved into NCW. A key tenet of *Joint Vision 2020* is that information superiority will enable decision dominance, new Joint operational concepts, and a decisive advantage over future adversaries. To create and leverage information superiority, it is foreseen that, under some circumstances, a mix of command, control, communications, computers, intelligence, surveillance, and reconnaissance (C⁴ISR) capabilities would interoperate with weapon systems and forces on an end-to-end basis through a network-centric information environment to achieve significant improvements in awareness, shared awareness, and synchronization. The military is embarked on a series of analyses and experiments to improve its understanding of the potential of these NCW concepts.

The Assistant Secretary of Defense for Networks and Information Integration (ASD NII), through the Command and Control Research Program, asked RAND to help develop methods and tools that could improve the assessment of C⁴ISR capabilities and processes to the achievement of NCW concepts, including awareness, shared awareness, and synchronization. In response to this request, the RAND

Corporation has been participating in the Information Superiority Metrics Working Group, under the auspices of ASD NII. The group's purpose is to describe key concepts and related metrics that are necessary to explore part of the proposed NCW value chain—from information quality through awareness, shared awareness, collaboration, and synchronization, to force effectiveness and mission outcome. This report presents a methodology—including metrics, formulas for generating metrics, and transfer functions for generating dependencies between metrics—for measuring the quality of information and its influence on the degree of shared situational awareness.

This research was conducted within the Acquisition and Technology Policy Center of RAND's National Defense Research Institute (NDRI). NDRI is a federally funded research and development center sponsored by the Office of the Secretary of Defense, the Joint Staff, the unified commands, and the defense agencies.

CONTENTS

FIGURES

TABLES

The military is formulating new visions, strategies, and concepts that capitalize on emerging information-age technologies to provide its warfighters with significantly improved capabilities to meet the national security challenges of the 21st century. New, networked C4ISR capabilities promise information superiority and decision dominance that will enhance the quality and speed of command and enable revolutionary warfighting concepts. Assessing the contribution of C4ISR toward achieving an NCW capability is a major challenge for the Department of Defense (DoD). Much like the development of a new branch of science, this requires defining concepts, metrics, hypotheses, and analytical methodologies that can be used to focus research efforts, identify and compare alternatives, and measure progress.

INTRODUCTION

An important first step is to improve our understanding of how improved C4ISR capabilities and related changes in command control processes contribute to the achievement of core information-superiority concepts, such as situational awareness, shared situational awareness, and synchronization. Establishing a quantifiable link between improved C4ISR capabilities and combat outcomes has been extremely elusive and is therefore a major challenge. In this work, therefore, we develop a *mathematical framework* that can facilitate the development of alternative measures of performance and associated metrics that assess the contribution of information quality and team collaboration on shared situational awareness. The emphasis is on the development of the framework.

The research reported here builds on the work of the ASD NII Information Superiority Metrics Working Group. This body has developed working definitions, specific characteristics and attributes of key concepts, and the relationships among them that are needed to measure the degree to which information-superiority concepts are realized and their influence on the conduct and effectiveness of military operations. The research is also consistent with the NCW Conceptual Framework, which DoD's Office of Force Transformation and ASD NII are developing jointly. The NCW Conceptual Framework is an assessment tool that includes measures, general forms for metrics, and relationships between the measures and metrics. It contains a large number of measures related to the complete array of concepts associated with NCW, ranging from networking hardware through decisionmaking capabilities and synchronization of actions. The group's metrics, and this report's scope, are largely limited to the information and awareness components of the NCW Conceptual Framework, and explore these components in more detail than does the framework.

We begin by defining a reference model for discussing such issues in terms of three domains: that of ground truth (the *physical* domain); that of sensed information (the *information* domain); and that in which individual situational awareness, shared situational awareness, collaboration, and decisionmaking occur (the *cognitive* domain). The C^{4}ISR process is seen as extracting data from ground truth and processing the data in the information domain to produce a common relevant operating picture (CROP). The quality of the CROP and the quality of team collaboration combine to heighten (or degrade) shared situational awareness in the cognitive domain.

THE ANALYTIC FRAMEWORK

The objective of this research is to develop a quantitative methodology that allows us to link improvements in C^{4}ISR capabilities to their effects on combat outcomes. For this first effort, we have confined our work to assessing the effects of data-collection and information-fusion processes, and the dissemination of the fused CROP on individual situational awareness and, through the collaboration process, on shared situational awareness.

Figure S.1, the C⁴ISR Information Superiority Reference Model, describes the activities associated with the above processes. This model envisions the three "domains" extending from the battlefield environment to cognitive awareness of the battlefield situation and decision.

This report uses a generic C⁴ISR architecture to build a model representing the contributions of these processes. The architecture can be thought of as a six-stage process that comprises the following:

0. acceptance of the existence of physical ground truth, restricted here to physical battlespace entities and their attributes (the initial state)

1. sensing of ground truth by an array of network sensors

2. fusion of sensor data by a centralized set of fusion facilities

3. distribution of resulting information (the CROP) to the users over a potentially noisy and unreliable network

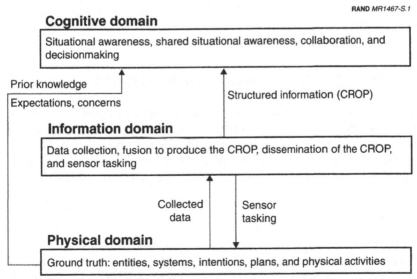

RAND *MR1467-S.1*

NOTE: The activities depicted in each of the domain "boxes" may not be complete. We focus on those activities pertinent to our research.

Figure S.1—The Information Superiority Reference Model

4. individual interpretation of the CROP, with the quality of the interpretation depending on the user's skills and abilities

5. collaboration to improve interpretation of the CROP, with the quality of the interpretation based on individual and group characteristics.

The value of the collection, fusion, dissemination, interpretation, and collaboration processes to combat operations within the above generic C^4ISR architecture is described through the several transformation functions, as shown in Figure S.2. The development of a quantitative framework is based on these transformations.

Enemy battlefield entities (units and weapon systems) are described in terms of their features or characteristics; hence, the quality of the information concerning the entities is an assessment of how well the C^4ISR system estimates the features of the collected set of enemy units in the battlespace. A conditional product form model is used to measure the effects of the NCW value chain transformations on the information-domain measures (quality of sensor information, quality of CROP, quality of shared CROP), and a more-general functional model measures the effects on the cognitive domain measures (situational awareness, shared situational awareness).

THE PHYSICAL AND INFORMATION DOMAINS

We applied the methodology to the measures in the physical and information domains. The feature matrix, $F = [F_1, F_2, \ldots, F_m]$, is a set of vectors, F_i, each of which represents the relevant physical characteristics of the enemy. In the physical domain, F_0 is a feature matrix representing the physical ground truth features of all enemy units.

Sensor Metrics

Using F_0 as an input, we first developed metrics formulas for the quality of sensor information, which is equivalent to the NCW Conceptual Framework's quality of organic information measure. Of the attributes the framework defined for the quality of organic information, we provide metrics for three: completeness, correctness, and currency.

RAND *MR1467-S.2*

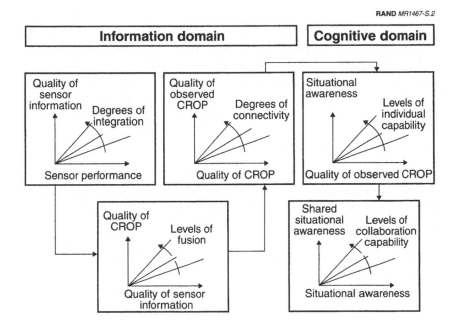

Figure S.2—The Information Superiority Value Chain

Completeness. We examined three aspects of completeness: the number of enemy units detected, the features reported for the units detected, and the sensor suite coverage area. For sensor information to be complete, all features of all units in the relevant ground truth must be known, and the entire area of operations must be under sensor observation. The suggested completeness metric has two components, both of which are between 0 and 1: c_1 is the fraction of enemy units detected (as specified in F_0), and c_2 is the fraction of the area of operations covered. We then have the following *transfer function* that, using F_0 as an input, combines these two components to produce a 0–1 completeness metric: $Q_{com}(F_1|F_0) = c_1(1-e^{-c_2})$. Here, F_1 is the CROP as detected by the sensors.[1]

Correctness. The metrics we suggest for correctness either support controlled experiments or support actual operations (in which analysts can only approximate ground truth from sensor inputs). In

---

[1] The body of the report presents rationales for all the metrics' functional forms.

either case, correctness is taken to mean the degree to which the true target features approximate their ground-truth values. Estimation theory is one way to assess the deviation from ground truth for controlled experiments. Since an unbiased estimator of a parameter is one whose expected value matches the true parameter, the difference between the estimate and the known ground truth appears to be a suitable metric to measure correctness. In general, if A is a measure of nearness, then $Q_{cor}(F_1|F_0) = e^{-A}$ is the transfer function we used to map A to a 0–1 metric.

Assessing correctness in support of operations implies that ground truth is not known. In this case, we cluster the detections geographically using a pattern-classification technique and then calculate the variance within the cluster. For a location estimate, the variance is expressed in terms of a covariance matrix. The determinant of that matrix is a measure of precision and therefore a measure of correctness. The determinant is $p = S^4$, where S^2 is the sample variance in both the x and y directions. $Q_{cor}(F_1|F_0) = e^{-p}$ is the transfer function we used to produce a 0–1 correctness metric.

Fusion Metrics

In the architecture we present here, the sensors transmit their readings to a series of fusion facilities, each of which focuses on a single intelligence discipline. Each facility submits its fused reports to a single central fusion facility, which combines the sensor inputs into a single, common, relevant picture of the battlespace: the *fused* CROP. This subsection develops metrics for the quality of the fused CROP, which is equivalent to part of the NCW Conceptual Framework's quality of individual information measure.[2] As noted, we assumed that the underlying network transmits the sensor readings to the fusion facilities perfectly.

Fusion includes the correlation and analysis of data inputs from supporting sensors and sources. Fusion occurs at several levels, from

---

[2]The quality of individual information measure is a multidimensional array measure, with the entries along one dimension corresponding to the quality of information seen by each individual. Further, one of those "individuals" is a user at the central fusion facility, who directly sees the fusion facility's output. This part measures the Quality of individual information as perceived by that user. The next section measures the quality of individual information perceived by users away from the central facility.

the simple combining of tracks and identity estimates to assessments of enemy intent. Our focus here is on the lower levels of fusion, which seek to improve the accuracy and completeness of the sensor reports on enemy units' features.

Completeness in the fusion subdomain focuses on the number of sensor-detected enemy units that have been *classified*, i.e., described in terms of their relevant features. The number of enemy-unit features the fusion facilities can classify depends on the architecture of the fusion suite, the degree of automation used, and the ability of the system to retask the sensors. The proposed formula for a 0–1 completeness metric is

$$Q_{com}\left(\mathbf{F}_2 \mid \mathbf{F}_0, \mathbf{F}_1\right) = \left[1 - \prod_{i=1}^{k}\left(1 - c_i\right)\right] c_c \, ,$$

where k is the number of subsidiary fusion facilities, c_i is the fraction of the detected enemy units that fusion facility i can classify per unit of time, c_c is the fraction the central processing facility can process, and $\mathbf{F}_2$ is the CROP after it has been through the fusion process.

Correctness in the fusion subdomain measures how close the fused estimate for each enemy unit feature is to ground truth. That is, how accurate are the classifications of the reported detections? One way we might address this problem is to examine the variance in the feature estimates for each reported unit. This results in the following formula for a 0–1 correctness metric:

$$Q_{cor,1}\left(\mathbf{F}_2 \mid \mathbf{F}_0, \mathbf{F}_1\right) = \sum_{i=1}^{n} w_i \sum_{j=1}^{p} \omega_j \, e^{-s_j} \, .$$

In this formulation, w_i and ω_j are weights. The former accounts for the relative importance of the reported enemy unit, and the second accounts for the relevant importance of the features being reported. The values of s_j are sample standard deviations for each of the p features for a given enemy unit, derived from the number of reports arriving on the unit. The second subscript on Q is used to distinguish this correctness transformation from the tracking metric discussed next.

An additional task is measuring how well we are able to track enemy units. The correctness of the tracks of enemy units can be measured in terms of the number of previous tracks that have been confirmed on the present scan and the number of new tracks initiated. The tracking portion of the correctness component of the transformation function is taken to be $Q_{cor,2}(F_2|F_0,F_1) = T$, where T is the fraction of the enemy units that correlate with previous tracks.

Combining the two correctness metrics using an importance weight, $0 \leq \omega \leq 1$ yields the following for the correctness component of the transformation function:

$$Q_{cor}\left(F_2 \mid F_0, F_1\right) = \omega W + \left(1 - \omega\right)T,$$

where $W = Q_{cor,1}(F_2|F_0,F_1)$.

Finally, an appropriate 0–1 metric for the currency attribute of quality of the fused CROP is $Q_{cur}(F_2|F_0,F_1) = e^{-t}$, where t is the total time required to update the fused CROP. This function emphasizes the importance of updating the fused CROP quickly.

Network Metrics

Following fusion, the architecture distributes the fused CROP to the force network's users, resulting in the *observed CROP*. Here, we provide metrics for the quality of the *observed CROP*, which is the remainder of this report's instantiation of the NCW Conceptual Framework's quality of individual information measure.[3] In these calculations, we allow the network to incur errors and delays in distributing the CROP. Thus, although we do not specifically incorporate the NCW Framework's Degree of Networking and Degree of Information "Shareability" metrics in this report, these metrics would directly influence the parameters of the functions used to generate the quality of the *observed CROP* metrics.

Thus, completeness here measures how well the communications network accommodates the transmission of relevant aspects of the CROP to each user. A metric for this measure is the probability that all users will receive the CROP. This is an assessment of the network's

---

[3]This section describes how to calculate the quality of individual information metrics for those users not at the central fusion facility, who must receive the CROP over the network.

reliability in terms of its robustness. The resulting completeness metric has the following formula:

$$Q_{com}\left(F_3 \mid F_0, F_1, F_2\right) = \prod_{i=1}^{k} p_i \, .$$

In this formulation, k is the number of users of the CROP, and p_i represents the probability that user i will receive the CROP.

Network correctness is an assessment of the likelihood that CROP users receive the distributed information without degradation. One way to measure this is to use the probability of correct message receipt (PCMR). The PCMR is a conditional probability that the message sent will be the message received. The probability that user i will receive the CROP (or a portion of it) as transmitted is $P_i(F_3, F_2) = P(F_2)P_i(F_3|F_2)$, where $P_i(F_3|F_2) = p$. We therefore get the following PCMR for user i:

$$\text{PCMR}_i = P_i\left(F_3, F_2\right) = P\left(F_2\right)p_i \, ,$$

where $P(F_2)$ is the probability that a user receives the CROP without error, given that the user receives the CROP. Therefore, our formula for a 0–1 metric for correctness is

$$Q_{cor}\left(F_3 \mid F_0, F_1, F_2\right) = \prod_{i=1}^{k} \text{PCMR}_i \, .$$

The end-to-end time required to transmit the CROP from the central fusion facility to the users serves as a measure of network currency. One way to determine this is to calculate the average of all paths from the source to the user. The overall average network transmission delay, then, is taken to be the average of these times, $\bar{t}$, so that a 0–1 metric for currency is

$$Q_{cur}\left(F_3 \mid F_0, F_1, F_2\right) = e^{-\bar{t}} \, .$$

Shared Information

Shared information is an essential ingredient to ensure effective collaboration. Recall that the CROP users receive is the *observed* CROP. Matrix F_2 represents the *fused* CROP. Each user's observed CROP is a

subset of the fused CROP. The overlap among these subsets constitutes the information shared among the users. Information not in the
overlap has the potential to be shared through the process of collaboration. The ability to collaborate therefore has the potential to
increase the amount of information shared among the users, thus
contributing to shared situational awareness.

Since "shared information" applies to subsets of the observed CROP,
the quality measures for Quality of Shared Information are equivalent to those for the Quality of the Observed CROP. A new attribute,
however, is the *extent* to which the observed CROP is shared. The
body of the report discusses various set-theoretic metrics for determining the extent of information sharing.

THE COGNITIVE DOMAIN

In the information domain, the data collected on the physical
domain are processed and disseminated to friendly users. In the
cognitive domain, the products of the information domain are used
to take decisions. The mental processes that transform CROP into a
decision and a subsequent action depend on a range of factors, a few
of which are psychological. The cognitive processes that transform
the CROP into a decision and subsequent action must be described
for participants in the decision process, both as individuals and as
interacting, collaborating members of a decisionmaking team. In this
report, we restrict our attention in the cognitive domain to how well
users can assess the situation presented to them through the
observed CROP. With respect to the NCW Conceptual Framework,
we restrict our attention to the Individual and Shared Awareness
measures, which are subsets of the framework's Individual and
Shared Sense-Making measures, respectively.

Modeling Individual Situational Awareness

Several factors influence what it will take for an individual decisionmaker to correctly assess the situation presented to him. Among
these is the *quality of the information* presented. This metric assesses
*the degree to which the decisionmaker is aware of the situation facing
him*, emphasizes the use of the individual components of the CROP,
and includes a reference to the *ability* of the individual decision-

maker. It is interpreted to be the fraction of the observed CROP the decisionmaker realizes.

We developed an agent representation of a decisionmaker using combinations of capability attributes (education and training, experience) and defined two discrete points for each attribute. From this, we produced four decision agents possessing these attributes at one of the two points. The agents suggest a functional relationship in which the dependent variable is "degree of awareness" and the independent variables are information quality measures (completeness in this case).

The end result of this process is an explicit relationship between the quality of the observed CROP and the ability of the decisionmaker.

Modeling Shared Situational Awareness

To describe *shared situational awareness*, we augmented the individual shared awareness model by representing the complex interactions in situations involving more than one individual. The metric we chose for this is *the fraction of fused feature vectors in the observed CROP that members of a team realize similarly, whether or not they collaborate*. This metric emphasizes the importance of individual situational awareness and allows agreement to exist even when individual decisionmakers have not collaborated.

We hypothesized, however, that *when collaboration is used, it is critical for determining shared situational awareness*. We focused on assessing the important attributes that affect teams that *do* collaborate and therefore have either positive or negative effects on the degree of shared awareness.[4]

One ingredient of the shared situational awareness process is the concept of a *common ground*. For our purposes, this term refers to the knowledge, beliefs, and suppositions that team members believe they share. During a team activity, therefore, common ground accumulates among team members.

---

[4]Note that these attributes, and the effectiveness of collaboration in general, are part of the NCW Conceptual Framework's Quality of Interactions measure.

We further hypothesized that, *to be effective, collaboration requires both the development of common ground among collaborators and familiarity with the capabilities of other collaborators.* Common ground does not develop instantaneously when there is collaboration; there is a period of "initial calibration" during which participants "tune in" to each other and move from a state of common sense to states of common opinion and common knowledge.

A structural model for defining and analyzing this phenomenon is a *transactive memory system,* defined as a set of individual memory systems in combination with the communication that takes place between individuals. It is concerned with the prediction of group (and individual) behavior through an understanding of how groups process and structure information.

Information can be stored and retrieved internally by an individual according to the individual's encoding, storage, and retrieval processes. If an individual stores information externally, the storage and retrieval process must also include the location of the information. If externally stored information resides in another person, a transactive memory system exists. Individuals can be assigned as information stores because of their personal expertise or through circumstantial knowledge responsibility. Each individual participating in the transactive memory has a set of memory components. These memory components capture the key elements of the collaboration. They represent information that some individuals store externally in other individuals and some individuals retain on behalf of other individuals in the transactive memory system. There can be direct links between an individual and the retrieval of a memory item and there can be indirect links that take "hops" through the transactive memory system until the memory item is accessed.

As participants develop stronger relationships with other participants through repeated or continued team interaction, the links between the participants become stronger. This suggests a second common ground hypothesis: *The completeness of the system for recording and retrieving information depends on how frequently the team has recently collaborated.* This concept is referred to as "team hardness."

A time-dependent functional model for team hardness is $0 \le TM(T) \le 1$, where $TM(T)$ is a function whose values are between 0 and 1, t represents the time elapsed since the start of the operation, and τ represents the length of time the team has been training or operating together, and $T = \tau + t$.

Consensus plays a central role in developing a transactive memory system. It is the majority *opinion* of a team arrived at through active collaboration. Its definition *implies the existence* of shared situational awareness. Noting that not all collaborating individuals have to agree before a decision and subsequent action can take place, we are interested in a measure of the degree of consensus. We hypothesized that *the degree of consensus can be estimated by the number of pairwise combinations of collaborating individuals who interpret feature vectors similarly.*

Models of shared situational awareness integrate the modeling proposed earlier. First, we placed the individual in a team and measured his situational awareness in a team setting. Note that this is not the same as team awareness but is rather the effect of team dynamics on an individual member of a collaborative decisionmaking process. The contribution is essentially derivative of the transactional memory function and, therefore, team hardness. Second, we addressed the consensus that develops among collaborating individuals and its effects on the team's shared situational awareness. Finally, we accounted for the diversity of decision-agent capabilities among the collaborators that results in our composite model for the degree of shared situational awareness.

FUTURE DIRECTIONS

As suggested, this work is clearly incomplete. We have described a mathematical framework that might be used to develop detailed mathematical quantities that represent what are generally considered qualitative concepts. In some cases, data may exist in the military C^4ISR community to confirm or disconfirm both the process and any of our examples. In these cases, locating and assessing the data are required. Where data do not exist, further experimentation or historical analysis will be required.

Much remains to be done in the cognitive domain. The relationship between information quality and situational awareness is the first step in the decisionmaking process. Further work is needed to codify the relationship between situational awareness and the ability of the decisionmaker to make inferences from the CROP—that is, his understanding of the situation.

Several techniques might be used to advance our knowledge in this important area. Among these are the following:

- **Data fitting.** We can use existing data either to confirm the validity of the relationships suggested in this work or to suggest different relationships.

- **Experimentation.** Experiments might provide additional insights about the relationships between information quality and awareness. For example, we could select decisionmakers that have various combinations of awareness characteristics. The degree to which they are able to realize enemy intent from what is presented, then, is an indicator of their level of awareness.

- **Decision and action.** The link between awareness and decision needs to be established. The level of awareness affects the ability to understand, i.e., to draw inferences about the CROP, such as enemy intent. The inferences, in turn, affect the decision to be taken and therefore the subsequent actions ordered.

- **Historical analyses.** Analysis of past battles is an important source of insight into the value of information. Considerable data is available from various sources that can provide insights into the relationship between the quality of information and the level of awareness.

- **Gaming.** It is also possible to use game theory to illustrate certain effects of information imbalances between two opponents. In a two-sided game, each side strives to obtain high-quality information. At the same time, each side attempts to ensure that the opponent's information is of low quality. Several pairings of players with varying awareness characteristics might then be played against the various information-quality levels. In this way, a link is established from information quality to awareness to decisions and, finally, to outcome.

ACKNOWLEDGMENTS

In conducting the research leading to this report, the authors consulted several RAND colleagues and government officials. We specifically acknowledge the guidance provided by David Alberts, Director of Research for ASD NII. Within RAND, Daniel Gonzales, Richard Marken, Thomas Sullivan, and Jonathan Mitchell made significant contributions to the quality of the report. Finally, the authors are grateful to the careful reviews provided by Richard Hayes, President of Evidence Based Research Inc., and RAND colleagues Paul Davis and John Hollywood. Their comments and suggestions greatly strengthened the quality of this report.

ABBREVIATIONS

AO	area of operations
ASD NII	Assistant Secretary of Defense for Networks and Information Integration
C^{4}ISR	command, control, communications, computers, intelligence, surveillance, and reconnaissance
CEC	cooperative engagement capability
COMINT	communications intelligence
COP	common operational picture
CPM	critical path method
CROP	common relevant operating picture
DoD	Department of Defense
ELINT	electronic intelligence
GMTI	ground moving target indicator
GUI	graphical user interface
IMINT	imagery intelligence
INT	intelligence discipline
IPB	information preparation of the battlefield
ISMWG	Information Superiority Metrics Working Group
MTI	moving target indicator

NCW	network-centric warfare
PCMR	probability of correct message receipt
ROF	ring of fire
SAR	synthetic aperture radar
SNR	signal-to-noise ratio
STARS	Surveillance Target Advance Radar System
UTM	Universal Transverse Mercator

actual common ground	The common ground a group has been confirmed to share
actual common knowledge	The common knowledge a group has been confirmed to share
assumed common ground	The common ground a group is believed to share
assumed common knowledge	The common knowledge a group is believed to share
awareness	A realization of the current situation
belief	A proposition an individual would assent to if given ample opportunity to reflect
cognition	The ability of a human to derive special or certain knowledge from an information source
collaboration	A process in which two or more people actively share information while working together toward a common goal
common ground	The knowledge, beliefs, and suppositions participants believe they share about the joint activity they are performing
common information	Information a group of users possesses

common knowledge	True common opinion
common opinion	A group proposition that each member believes, believes each member believes, believes each member believes each member believes, ad infinitum
completeness	The degree to which the information is free of gaps
consistency	The extent to which information is in agreement with related or prior information
correctness	The degree to which information agrees with ground truth
currency	The situation *independent* time required for the C^{4}ISR system to produce and distribute a CROP
data	Any representation to which meaning might be assigned
exploratory analysis	Evaluating the individual impact of several alternatives on the outcome of a process
feature	A prominent part or characteristic of the combat situation
fusion	Combining information from disparate sources and sensors to form a CROP
information	Data that have been processed in some way
information quality	The inherent "goodness" of information; information quality is *situation independent*
information superiority	The ability to collect, process, and disseminate information as needed; anticipate the changes in the enemy's information needs; and deny the enemy the ability to do the same

Information Superiority Reference Model	A depiction of the physical, information, and cognitive domains
information value	Information that is useful to the decision to be taken; valuable information is *situation dependent*
knowledge	Accumulated and processed information wherein conclusions are drawn from patterns
measure	A basis or standard of comparison
metric	Mathematical formulas used to evaluate the differences among alternatives
multisensor integration	Connecting sensors in such a way as to increase their collective ability to detect and identify unit/targets
shared information	Information available to a group of users
shared situational awareness	The ability of a decisionmaking team to share realizations about the current situation
situational awareness	Realization of the current situation based on the observed CROP
timeliness	The situation-dependent degree to which information is available when needed
transactive memory	A set of individual memory systems in combination with the communication that takes place between individuals
understanding	The ability of humans to draw inferences about the possible consequences of a situation

Chapter One

INTRODUCTION

> It is the quality of our work that will please God and not the quantity.
>
> —*Mahatma Gandhi*

The military is formulating new visions, strategies, and concepts that capitalize on emerging information-age technologies to provide its warfighters with significantly improved capabilities to meet the national security challenges of the 21st century. New, networked command, control, communications, computers, intelligence, surveillance, and reconnaissance (C⁴ISR) capabilities promise information superiority and decision dominance that will enhance the quality and speed of command and enable revolutionary warfighting concepts. Assessing the contribution of C⁴ISR toward achieving a network-centric warfare (NCW) capability is a major challenge for the Department of Defense (DoD) because of the multiplicity of interacting factors and the lack of understanding of the fundamentals associated with information-superiority concepts. DoD has embarked on a journey of exploration to discover how to create and leverage information superiority by characterizing the conditions under which it can be achieved and under which a competitive advantage can be gained. Much as in the development of a new branch of science, this requires defining concepts, metrics, hypotheses, and analytical methodologies that can be used to focus research efforts, identify and compare alternatives, and measure progress.

In response to this need, DoD's Office of Force Transformation, in conjunction with the Office of the Assistant Secretary of Defense for Networks and Information Integration (ASD NII), has been evolving a conceptual NCW framework for assessment that includes measures, general forms for metrics, and relationships between the measures and metrics. The NCW Conceptual Framework is intended for exploratory analysis of potential C⁴ISR architectures and for guiding C⁴ISR experiments in terms of identifying the measures for which to collect experimental data.[1]

The technical, quantitative application of this framework requires developing mathematical models that reflect the performance of the enabling C⁴ISR processes, architectures, and capabilities, as well as developing a hypothesis identifying the key variables to measure, and their effects. Consequently, this report describes a methodology for developing and linking relevant mathematical models. It provides mathematical models corresponding to specific C⁴ISR architectures and illuminates the data needed to apply the NCW Conceptual Framework in a specific context.

RESEARCH OBJECTIVES

An important first step is to develop a better understanding of how improved C⁴ISR capabilities and related changes in command and control processes contribute to the achievement of such core NCW concepts as situational awareness, shared situational awareness, and synchronization. The next step is to assess how well the core concepts affect command and control operational concepts and, in turn, the success of military operations. Establishing a quantifiable link between improved C⁴ISR capabilities and combat outcomes has been extremely elusive and is therefore a major challenge.

The primary objective of our work was to develop a clearly articulated *mathematical framework* that would explore how certain major factors would affect the hypotheses related to NCW. The scope of the research is restricted to the NCW Conceptual Framework's key measures related to *information superiority*, which Joint Publication 1-02

---

[1]The most recent public document on the NCW Conceptual Framework is Signori et al. (2002).

defines as "[t]hat degree of dominance in the information domain which permits the conduct of operations without effective opposition" (DoD, 2003, p. 255).

In particular, we examined how the quality of information (individual and shared) a C⁴ISR network provides affects the quality of situational awareness (also individual and shared). In the NCW Conceptual Framework, the Quality of Information and Quality of Awareness are a vector of submeasures. The research focuses on a subset of these submeasures, notably completeness, correctness, and currency.

The modeling approach was to study a chain of information quality and awareness quality metrics in three domains: physical, information, and cognitive. The models assumed that a specific type of C⁴ISR architecture is being employed, in which an array of sensors transmit data about the battlespace to a central fusion facility that distributes a resulting common relevant operating picture (CROP) to force members. The force members then mentally interpret the CROP they receive (creating awareness) and collaborate with other force members to improve everyone's awareness.

This report emphasizes the development of the mathematical framework. Although we include several measures and mathematically defined metrics, they should not be considered the results of settled research. They are necessarily simple and illustrative rather than general and widely applicable.

The quantitative methodology and illustrative mathematical representations here relate to force-on-force combat operations as opposed to a broader spectrum of operation that includes aid to civil authorities, humanitarian relief, and peacekeeping operations. However, aspects of the approach provide a basis for informed dialog that, in our view, will eventually lead to more-comprehensive and better-validated capabilities to quantitatively explore the influence of improved C⁴ISR systems and processes on operational outcome.

ANALYTICAL FRAMEWORK

In addition to the NCW Conceptual Framework, the research reported here builds on the work of the ASD NII Information Superiority Metrics Working Group (ISMWG). This body has developed working definitions, specific characteristics and attributes of key concepts, and the relationships among them that are needed to measure the degree to which information superiority concepts are realized and their impact on the conduct and effectiveness of military operations. Such an endeavor requires a common language and a set of integrated hypotheses, as well as the metrics, instruments, and tools to collect and analyze data, such as those suggested here.

We begin by defining a reference model for discussing such issues in terms of three domains: that of ground truth (the *physical* domain); that of sensed information (the *information* domain); and that in which individual situational awareness, shared situational awareness, collaboration, and decisionmaking occur (the *cognitive* domain). The C⁴ISR process is seen as extracting data from ground truth and processing the data in the information domain to produce a CROP. The quality of the CROP combines with the quality of team collaboration to heighten (or degrade) shared situational awareness in the cognitive domain. The ground truth data, obtained from the collection process in the information domain, is *transformed* into a CROP that contributes to situational awareness in the cognitive domain. The transformations are processes that include data collection and processing, data fusion, and information dissemination. These processes are not discussed in this report. We assume that they are performed and focus instead on the *quality* of the information and products they generate.[2]

LIMITATIONS

This report focuses on the collection of data; the processing of collected data to produce the CROP; the dissemination of the CROP from the various fusion facilities to the ultimate users; the experience of the decisionmaking team; the quality of its collaboration; and, finally, how all this affects shared situational awareness. In developing mathematical representations of information quality and the effects of collaboration, we focused exclusively on battlefield *entities*—individual units and weapon systems. The possible inferences

---

[2]For a complete discussion of C⁴ISR information processing algorithms, see Perry and Sullivan (1999).

to be drawn from the patterns these entities exhibit, although very important, are beyond the scope of this initial work.

We also do not treat several important aspects of the decisionmaking cycle, such as *decisionmaking* and *synchronization.* The relationship between awareness and understanding and how understanding affects decision and action are also not addressed. Nevertheless, this document presents a new and important methodology for assessing the quality of information and intelligence, surveillance, and reconnaissance processes in general, as well as some of the more psychological aspects of the decisionmaking process. It is based on sound mathematical concepts and hypotheses from the literature and therefore provides a foundation for further inquiry. The illustrative mathematical relationships, however, will require verification and refinement through experimental and operational data.

Finally, with respect to the NCW Conceptual Framework, this report is restricted to the information and awareness measures. The report does not attempt to show models that traverse all the measures in the framework.

ORGANIZATION OF THIS REPORT

Chapter Two outlines the analytic framework we adopted to assess the effects of information quality and team collaboration on shared situational awareness and, eventually, decision and execution. We begin by describing the C⁴ISR Information Superiority Reference Model as the underlying construct that describes the C⁴ISR process. Chapter Three focuses on a mathematical framework for assessing the contributions of both the physical and information domains to information quality. Chapter Four turns to the more-psychological aspects of decisionmaking and assesses the contributions of the cognitive domain to shared situational awareness. Finally, Chapter Five addresses future work. Three appendices are included: Appendix A lists the definitions of important terms; Appendix B records a few example mathematical representations for the metrics developed in the main text; and Appendix C describes a spreadsheet model used to illustrate the framework using the example metrics listed in Appendix B.

Chapter Two

THE ANALYTIC FRAMEWORK

> A decision is an action an executive must take when he has information so incomplete that the answer does not suggest itself.
>
> —*Arthur William Radford*

This chapter outlines the analytic framework we adopted to assess the effects of information quality and team collaboration on shared situational awareness and, eventually, decision and execution. We begin by describing the underlying construct that describes the C⁴ISR process we model here: the C⁴ISR Information Superiority Reference Model. The quality of the information flowing through what we call "the information value chain" is transformed to produce an overall assessment of the quality of the CROP generated. This, in turn, affects individual situational awareness and, subsequently, shared situational awareness through the process of team collaboration.

THE C⁴ISR INFORMATION SUPERIORITY REFERENCE MODEL

The C⁴ISR Information Superiority Reference Model (Figure 2.1) is a representation and extension of the portion of the NCW Conceptual Framework dealing with information superiority issues (in particular, the information and awareness measures). It represents the activities associated with collecting data; processing information; discerning enemy intent, plans, and physical activities (or lack thereof); selecting a course of action; and monitoring its execution. The model consists of three "domains" that extend from the battlefield environment to cognitive awareness of the battlefield situation and decision. The measures and metrics proposed here, although not all-inclusive, are presented within this context.

Both sides in a conflict generally have different perceptions of a single reality, referred to as the *situation*. Figure 2.1 depicts how the three domains contribute to this perception. The major activities performed in each of the domains are listed in its box. The physical domain is where reality or ground truth resides.

In addition to physical objects (*entities*)—weapon systems, terrain features, sensors, etc.—the physical domain also contains intangibles, such as enemy intent, plans, and current and projected activities. A complete assessment of the situation will contain estimates about each. As mentioned earlier, however, we will focus on entities only, thus reserving the others for future research.

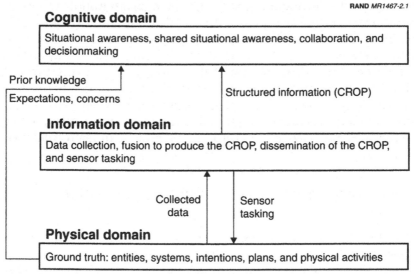

NOTE: The activities depicted in each of the domain "boxes" may not be complete. We focus on those activities pertinent to our research.

Figure 2.1—The Information Superiority Reference Model

In the information domain, data are extracted from the physical domain and processed to form structured information in the form of a CROP. Three primary functions are performed in the information domain: collecting data through the use of sensors and sources, including tasking sensors to close gaps in the data; processing the data through the fusion process to produce the CROP; and disseminating relevant parts of the CROP to friendly units. The last step contributes to the collaboration process in the cognitive domain, in which the shared CROP is transformed into a shared awareness of the current and future situations that can be used to gain understanding of threats and opportunities, as well as the subsequent decisionmaking regarding an appropriate course of action.

Finally, the human activities associated with using the information available to form an estimate of the situation take place in the cognitive domain. To the extent that decisionmaking teams exist, they collaborate to form a level of situational awareness. In addition to the CROP produced in the information domain, individual team members and the decisionmaker may have prior information from such processes as the information preparation of the battlefield (IPB) available to support their deliberations. Finally, the decisionmaker is likely to have concerns and expectations about the performance of his own forces, as well as those of the enemy, that would color his assessment of the situation and therefore his decision. These are depicted as emanating directly from the physical domain.

THE C⁴ISR ARCHITECTURE

Employing the NCW Conceptual Framework and the Information Superiority Reference Model requires a C⁴ISR architecture. The architecture used in this report may be thought of as a linear process with six steps:

0. **Ground Truth.** The architecture begins with acceptance of the existence of physical ground truth. As the previous chapter noted, we restricted our consideration of ground truth to battlefield entities, such as individual units and weapon systems, and their physical attributes.
1. **Sensing.** The architecture first uses a set of sensors to detect the battlespace entities and their sensors.

2. **Fusion.** The sensors then transmit their data to fusion facilities. The fusion structure used in this report is centralized and has two stages. In the first stage, incoming sensor data are sent to one of k fusion facilities, each corresponding to a different intelligence discipline. In the second stage, the partially fused data from each discipline are forwarded to a central processing facility, which generates a single CROP. For the sake of simplicity, we assumed that the network connections transmitting data from the sensors to the fusion facilities are flawless.

3. **Distribution.** The central processing facility then transmits various versions of the CROP to the network's users. At this point, we allowed the modeling of loss of service, errors, and delays in the network.

4. **Individual Assessment.** Each user then attempts to interpret the CROP he has received to achieve some level of realization of the battlespace.

5. **Group Assessment.** The users then collaborate with each other in an attempt to improve their realization of ground truth in the battlespace. This report models the effectiveness of collaboration as a function of the skills of the users and the collaborative group as a whole but does not examine the affects of the network's communications tools on collaboration.

THE NCW VALUE CHAIN

In conjunction with the NCW Conceptual Framework and the Information Superiority Reference Model, a hierarchy of hypotheses regarding NCW concepts is emerging as the focus of experimentation and analysis. Central among these is the hypothesis that improved networked C⁴ISR capability will improve information quality and shared situational awareness, allow more-dynamic decisionmaking and agile force synchronization, and will ultimately increase force effectiveness and the likelihood of a successful operational outcome. The basis of the hypothesis is that improved networked C⁴ISR capability will add *value*.

Consequently, we used the reference model to establish the information superiority *value chain* and examined it quantitatively to determine which factors have the greatest payoff and the conditions under which benefits accrue. Figure 2.2 depicts the initial portion of this value chain, which represents the hypotheses from the NCW Conceptual Framework that are associated with the Information Superiority Reference Model and also applies directly to the C⁴ISR architecture discussed above.

In particular, the value chain represents the processes associated with the collection, fusion, and dissemination of information, as well as the degree to which the information contributes to shared situational awareness and, ultimately, to the decision rendered, focusing on the value these processes add (or subtract) at each step. Examination of the value chain will require measuring key aspects of the end-to-end process—not only the quality of information and degree of shared situational awareness but also the functional performance of the people and systems that transform these attributes along the way. The effects of such functions will vary with the degree of integration in the sensor suite; the level of fusion available at the fusion centers; the connectivity of the communications network; the capability of the decisionmakers; and the type of command and control process, including the degree of collaboration.

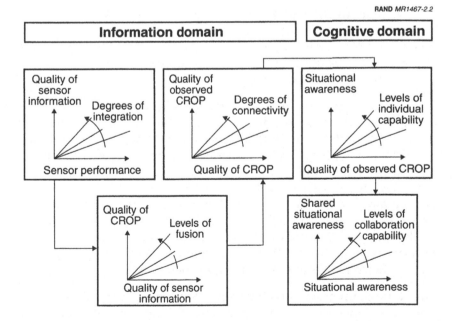

Figure 2.2—The Information Superiority Value Chain

The five graphs in Figure 2.2 depict illustrative cascading transfer functions that successively map performance in one part of the value chain, as well as exogenous inputs, to performance in the next part of the value chain. The emphasis is on the quality of the processes, not on the processes themselves—on "how well the sensor suite performs" and not on "what it performs." The first transfer function addresses the effect of sensor performance on the quality of the information produced. The degree and nature of the multisensor integration can be altered parametrically to obtain the shifts in Quality of Sensor Information depicted in the figure. The quality of the information obtained from the sensor suite, as well as the level of fusion (an exogenous input), contributes to the Quality of the *fused* CROP. In the figure, the levels of fusion refer to the extent to which the fusion facility is able to produce not only information on battlefield entities but also assessments of enemy intent. The *observed CROP* depicted in the next chart in the figure refers to the CROP distributed to the ultimate users, generally members of a collaborative team. The Quality of the Observed CROP, therefore, depends on the communications network that facilitates its transmission.

The versions of the CROP distributed to various users may vary. Thus, we are interested in the similarities between the users' observed CROPs. Although not directly shown on the figure, the Quality of the Shared CROP measure describes the consistency between users' observed CROPs.

In the next chart in the figure, we leave the information domain for the cognitive domain by depicting the relationship between the observed CROP and individual awareness. Note that situational awareness replaces quality of information as the dependent variable. This is the beginning of the team collaboration process, and its effectiveness will vary with the ability of the individual team member. Finally, the degree to which the individual team members are aware of the situation affects shared situational awareness. The quality of the collaboration depends on several factors, such as the experience of the team, how long its members have worked together, and the personalities and position of its members.

The value chain described above represents an application of part of the NCW Conceptual Framework. The differences and similarities are as follows:

- **Quality of Sensor Information** implements the framework's Quality of Organic Information measure but is restricted to sensor data.

- **Quality of the Fused CROP** partially implements the framework's Quality of Individual Information measure and assumes the existence of a centralized facility responsible for fusing the sensed data (see below); this measure captures the quality of the information a user at this central facility would see.

- **Quality of the Observed CROP** implements the rest of the framework's Quality of Individual Information measure. The transfer function for this measure incorporates the performance of a network in transmitting the CROP to various users. In the framework, network performance in transmitting particular pieces of information comes under the Degree of Information "Shareability" measure. Thus, the transfer function for Quality of the Observed CROP models the dependency in the framework between the shareability and quality of information.

- **Quality of the Shared CROP** implements the framework's Quality of Shared Information measure.

- **Quality of Situational Awareness** implements the Awareness submeasure of the framework's Quality of Individual Awareness measure; this report does not cover the second submeasure, Understanding.

A systematic analysis of the value chain helped us to understand the complexities associated with the end-to-end C4ISR process. A quantitative methodology that would account for links, key models, and parameters and that would permit examination of the highly nonlinear effects within the value chain would facilitate analysis. As an initial step, we proposed such a methodology for exploring the initial portion of the value chain as depicted in Figure 2.2, which we discuss below.

INFORMATION QUALITY

The NCW Conceptual Framework uses eight attributes to describe the quality of information, whether the information is organic, individual, or shared. The attributes are divided into two groups of four (see Table 2.1):

Table 2.1

Objective and Fitness Measures

Measure	Definition
Objective measures	
Correctness	The extent to which information is consistent with ground truth
Consistency	The extent to which information is in agreement with related or prior information
Currency	The age of the information
Precision	The level of measurement detail in an information item
Fitness measures	
Completeness	The extent to which information relevant to ground truth is collected ("relevant to ground truth" depends on the scenario)
Accuracy	The appropriateness of the precision of information to a particular use
Relevance	The proportion of information collected that is related to the task at hand
Timeliness	The extent to which the currency of information is suitable to its use

- *Objective measures* can be used regardless of the particular scenario

- *Fitness-for-use measures* require a particular scenario to be defined. Put another way, the objective measures are context-free, while the fitness for use measures are context dependent.

The three measures of information quality we selected from those listed in Table 2.1 were *completeness, correctness,* and *currency.* We also modified the definitions slightly for our use, as shown in Table 2.2. As noted, *completeness* is a fitness-for-use measure because its use requires determining what the "relevant ground truth" comprises. The other two measures are objective, although one can incorporate contextual information in defining metrics for them. For example, if, in a given situation, 200-meter accuracy is all that is needed, that can be made to equate to absolute correctness.

Table 2.2

Measures of Information Quality

Measure	Description
Completeness	The degree to which the information is free of gaps with respect to the relevant ground truth
Correctness	The degree to which the information agrees with ground truth
Currency	The time required to produce a CROP

Completeness

The definition of *completeness* in Table 2.2 is admittedly restrictive. This is because we confined our definition of *relevant* to a CROP that consists of information about enemy units in the area of operations (AO). In this context, having information about all relevant features of all enemy units (no gaps) means that the information is complete. In many scenarios, we can further restrict relevance to certain key features about the enemy units. Here, *relevant* implies that completeness depends on the situation, the command level, and the scale of the operation. For example, if we are interested in determining whether an enemy army is on the move, we may have complete information if we observe only a few vehicles moving. In this case, the pattern established (relevant entities on the move) is enough to assess the information as complete.

Correctness

Correctness, as defined in Table 2.2, compares what is observed, processed, and disseminated with the ground truth. This applies equally if the enemy is engaged in deception or not. As with completeness, correctness, in this context, focuses on the degree to which what is reported is close to truth. For example, if a unit is on the move and if the sensors and processing facilities produce a CROP that accurately records the location, speed, and direction of travel of the unit, the information is taken to be correct. However, we recognize that this says little about the enemy unit commander's intentions. Is this part of a feint? Is this a movement to contact or is it an administrative march? Answers to these questions are arrived at by examining patterns and drawing inferences from them, something we recognize requires further study.

Currency

Our use of the currency measure rather than timeliness indicates that we are only interested in how long it will take to get information to the users not whether it will arrive in time. Although the latter requires that we calculate the former, they are two distinct concepts. However, like the other measures, currency can be viewed in context. If we assume that faster is better, information that can be processed quickly has high quality. However, asking how fast is fast enough introduces context, which means we are really dealing with timeliness.

A QUANTITATIVE METHODOLOGY

A mathematical representation of the transformations depicted in Figure 2.2 must account for the changes in information quality and situational awareness associated with the processes depicted. In subsequent chapters and in Appendix B, we suggest some simple conceptual models for assessing information quality and situational awareness. The models incorporate metrics that measure the quality of products that the processes depicted in Figure 2.2 generate, as well as the effects of collaborative processes on situational awareness. These models can be linked using a basic quantitative methodology. The intent is to illustrate the methodology and provide a modeling baseline for further extension as data become available and as theory is fully incorporated, particularly where less tractable parameters are involved. To the extent that data are available or that more-detailed models exist, the developed models can be refined and tailored to the problem at hand. However, depending on the particular problem, other models will likely be more appropriate. Exploratory modeling and analysis provide tools for examining the effects of a variety of alternative models to account for a range of behavior or uncertainty associated with complex C^4ISR systems (see Bankes, 1993).

The methodology has three segments. The first quantifies key features of the real world; to begin with, these are confined to battlefield entities, such as units and weapon systems. As mentioned earlier, we have deferred the assessment of patterns that reveal enemy intent to future research. The second segment quantifies the quality of information as it transits the sensors, fusion centers, and distribution networks of the C⁴ISR infrastructure. The third quantifies the influence of the resulting quality of information on the degree of shared situational awareness through the collaboration process. In each case, we defined metrics that quantitatively characterize the key features associated with the situation, as well as the performance of the people and systems that process the information. In the case of the degree of shared situational awareness, we postulated that the transfer functions that reflect the effects of the quality of information would vary with a few dominant parameters associated with individual and group behavior. We chose these parameters on the basis of a brief review of the relevant literature.

For the quality of information the infrastructure produces, we developed operationally meaningful metrics so that they or their properties could be calculated using estimation theory. We used a different approach for assessing situational awareness. The experience of the members of the collaborative team and their effectiveness working as a team were the primary measures. In any case, however, the metrics were reduced to a number between 0 and 1. It seemed intuitive, therefore, to view the value chain as a series of *conditional* transfer processes, i.e., the quality of the information the fusion process produces is *conditioned* on the quality of the information the sensor suite produces. This allowed us to construct a series of transfer functions that compute numerical metrics by applying a conditional product-form model.

The system produces information about the enemy units and weapon systems (battlefield entities). Situational awareness is about the ability to base inferences about the enemy units and weapon systems on the observed CROP and other individual and team factors.[1] The information about an enemy unit is expressed in terms of unit *features*: prominent parts or characteristics of the unit. For example, one prominent feature might be the location of the unit; others would be its identity or combat strength. We collected all these fea-

---

[1] Chapter Four discusses,this more fully.

tures into a set and refer to it as the *feature vector*, denoted
$F = [f_1, f_2, \ldots f_n]^T$, where each f_i represents an individual feature.
Collecting all the feature vectors for all the enemy units in the battlespace produces a matrix: $\mathbf{F} = [F_1, F_2, \ldots, F_m]$. For our purposes, the
CROP comprises estimates of the features in this matrix, and the
degree to which it reflects ground truth is a measure of its quality.
Making inferences about the units then consists of the ability to "fill
in the gaps in the features."

The process Figure 2.2 depicts is sequential, consisting of five steps.
Let $\mathbf{F}_i$ be a matrix of the CROP's estimates of the feature vectors at the
end of the ith step. The quality of information associated with the
feature matrix, $\mathbf{F}$, at the output of the ith process ($i \geq 1$) is defined as
follows:

$$Q\left(\mathbf{F}_i, \mathbf{F}_{i-1}\right) = Q\left(\mathbf{F}_{i-1}\right) Q\left(\mathbf{F}_i \mid \mathbf{F}_{i-1}\right).$$

where $0 \leq Q(\mathbf{F}_{i-1}) \leq 1$ is the quality of information in $\mathbf{F}_{i-1}$; $Q(\mathbf{F}_i, \mathbf{F}_{i-1})$ is
the quality of the information at the output of the ith process, based
jointly on the estimates in the feature matrices in the last step and in
the current step; and $Q(\mathbf{F}_i|\mathbf{F}_{i-1})$ is the transfer function calculating a
partial value of $Q(\mathbf{F}_i)$ given a specific $\mathbf{F}_{i-1}$.

Note that $Q(\mathbf{F}_i)$ is a vector-valued function, with entries corresponding to a particular quality of information measure (completeness,
correctness, or currency). Thus, the product shown above is a scalar
(a term-by-term multiplication of elements), not a matrix, product.

A full expansion of the methodology results in the following chained-
product transformations:

$$Q\left(\mathbf{F}_0, \mathbf{F}_1, \mathbf{F}_2, \mathbf{F}_3\right) = Q\left(\mathbf{F}_0\right) Q\left(\mathbf{F}_1 \mid \mathbf{F}_0\right) Q\left(\mathbf{F}_2 \mid \mathbf{F}_0, \mathbf{F}_1\right) Q\left(\mathbf{F}_3 \mid \mathbf{F}_0, \mathbf{F}_1, \mathbf{F}_2\right)$$

$$A = h\left(Q\left(\mathbf{F}_0, \mathbf{F}_1, \mathbf{F}_2, \mathbf{F}_3\right), \mathbf{F}_4\right)$$

$$SA = g\left(A, \mathbf{F}_5\right),$$

where $Q(\mathbf{F}_0, \mathbf{F}_1, \mathbf{F}_2, \mathbf{F}_3)$ is the quality of information in the observed
CROP. Matrix $\mathbf{F}_0$ is ground truth; its quality is therefore taken to be 1:
$Q(\mathbf{F}_0) = 1$. A is the degree of individual awareness, based on the quality of the observed CROP, the expertise of the individual, and other

information available to him. From this point, the product-form model no longer applies, and the natures of the functions h and g are included in the development of the metrics in the cognitive domain. Finally, *SA* is the measure of shared situational awareness, which is dependent on individual awareness and the characteristics of the collaborating team.

SUMMING UP

The objective of this research is to develop a quantitative methodology that allows us to link improvements in C^4ISR capabilities to their effects on combat outcomes. For this first effort, we have confined our work to assessing the effects of data-collection and information-fusion processes and the dissemination of the fused CROP on individual situational awareness and, through the collaboration process, on shared situational awareness.

We use the Information Superiority Reference Model (Figure 2.1) to describe the activities associated with these processes. The model consists of three "domains" extending from the battlefield environment to cognitive awareness of the battlefield situation and decision. The transformation functions that the information superiority value chain (Figure 2.2) depicts describe the value of these processes to combat operations. The quantitative methodology is based on these transformations.

Enemy battlefield entities (units and weapon systems) are described in terms of their features or characteristics; hence, the quality of the information concerning the entities is an assessment of how well the C^4ISR system estimates the features of the collected set of enemy units in the battlespace. A conditional product-form model is used to measure the effects of the value chain transformations for information quality, while the effect of information quality on individual awareness and shared situational awareness is defined through a more general functional relationship. We defer the detailed discussion of the last set of functions to Chapter Four.

Chapter Three

THE PHYSICAL AND INFORMATION DOMAINS

> It is easier to perceive error than to find truth, for the former lies on the surface and is easily seen, while the latter lies in the depth, where few are willing to search for it.
>
> —*Johann Wolfgang von Goethe*

The physical and information domains are the sources of information used to inform decisions in the cognitive domain. We discuss them together in this chapter because of their close connection: The ground truth resident in the physical domain is approximated through data collection and processing in the information domain. The observed CROP is the approximation of the relevant portions of ground truth disseminated to friendly users.

THE PHYSICAL DOMAIN

The physical domain is the beginning and the end of the C⁴ISR cycle—both the subject of and object of decision. Chapter Two described this domain as consisting of the disposition of friendly and enemy forces in the AO, including friendly and enemy collection assets, fusion facilities, networks, and command and control facilities; the geospatial features of the region in which the forces are engaged; and such intangibles as enemy intent, plans, and current and projected activities. Before combat begins, information about ground truth is generally available through the IPB process. Information about some, but not all, elements of ground truth is perishable. In any case, the friendly commander's knowledge about all aspects of ground truth is uncertain. Part of the C⁴ISR system's function is to reduce that uncertainty as the operation unfolds. Once a decision has been reached and actions, including force *synchronization*, have been taken to ensure its implementation, the elements of the physical domain may be altered in a way that may not be known to the friendly commander. Thus the cycle begins anew.

Features

The physical domain's ground truth is described in terms of the features of the entities the CROP comprises. For our purposes, the CROP consists of information about enemy units only. Higher-level assessments that would include enemy intentions and plans, based on patterns and other information, are a next step. A *feature* is a prominent part or characteristic of a unit that is relevant to the friendly commander; Table 3.1 offers examples. Enough features must be collected to communicate to the commander the current estimate of the combat situation. Thus, the list of features does not necessarily describe all the units in the physical domain completely. To the list of features in the table, we could add the destination of the force, the activity it is currently engaged in, its echelon, etc.

The Relevant Ground Truth

Chapter Two characterized the relevant ground truth as consisting of a matrix, $\mathbf{F} = [F_1, F_2, \ldots, F_m]$, where each F_i represents a vector whose elements are the features of unit i. Our previous discussion noted

Table 3.1

Some Unit Features

Feature	Description
Location	The current placement of the unit in the battlespace, using x-y coordinates
Speed	Rate of advance to friendly positions, in kilometers per hour
Direction	Azimuth, in radians
Strength	A combat effectiveness score
Unit type	The predominant military organization type

that the features necessarily vary over time, for example, the location of a moving unit. This implies that matrix **F** is time dependent. Although we have made no notational adjustments to the matrix, it should be understood that it is dynamic. In addition, we track the status of estimates of **F** through domains by assigning subscripts to indicate where we are in the process. Accordingly, the matrix in the physical domain is designated F_0. In the physical domain, the quality of the information about F_0 is considered "perfect," in that it is ground truth, so that $Q(F_0) = 1$. Next, in the information domain, the quality of the information in the CROP reflects how well it represents the relevant ground truth, measured in terms of the completeness, correctness, and currency of the data provided by the sensor suite and how well the data have been processed and analyzed.

THE INFORMATION DOMAIN

Information collection, processing, and dissemination take place in the information domain, which is further subdivided into three subdomains: *sensor* (intelligence, surveillance, and reconnaissance), *fusion*, and *network*. These three constitute the main effort of the C^4ISR system and parallel the three information domain graphs depicted as part of the information superiority value chain in Figure 2.2. The sensors collect data about the relevant ground truth units from the physical domain, then forward it to fusion facilities, where it is processed, fused, and eventually shared with friendly users through a communications network. Figure 3.1 depicts the transformation equations from Chapter Two, showing the transformations from collection in the sensor subdomain to dissemination in the network subdomain. The end product is an observed CROP, a portion of which is available to each friendly user.

THE SENSOR SUBDOMAIN

Information about the physical domain originates with the sensors and information sources allocated or directed to the AO. The sensor array is generally referred to as the sensor suite. Its configuration will vary with the operational situation and the characteristics of the sensors. The output from the sensor domain is a set of feature vectors,

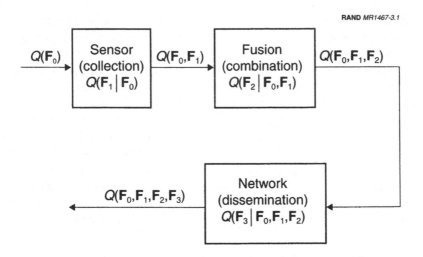

RAND *MR1467-3.1*

Figure 3.1—Information Domain Transformations

F_1, organized by discipline (such as imagery intelligence [IMINT], electronic intelligence [ELINT], or moving target indication [MTI]). The quality of this representation of the CROP, $Q(F_0, F_1)$, is dependent on the relevant ground truth, depicted as F_0 in Figure 3.1, and the performance of the sensor suite.

Following a short discussion of sensors and their characteristics, we suggest measures for the completeness, correctness, and currency of the sensor reports arriving at fusion facilities.

Sensors and Sources

Sensors are designed to detect objects, record images of designated areas, and estimate physical phenomena. They are capable of performing systematic surveillance and reconnaissance over large areas, subject to the existence of threats that may jeopardize sensor platform survival. Sensors may be capable of detecting types or classes of militarily relevant objects or targets or may detect whole classes of objects, such as moving vehicles. In general, sensor performance is a function of the environment (terrain, foliage, electromagnetic background noise, extraneous reflected sunlight or glint, etc.). For example, radar sensors that operate in the microwave band can detect only targets that have a radar cross section above some minimum threshold and that are in environments with signal-to-noise ratios above some minimum threshold.

Sources, on the other hand, may be covert and typically operate over much smaller areas. Sources include such things as human observations; very short range communications intercepts or surveillance; unattended covert devices that can be read out intermittently; and a priori knowledge about enemy force dispositions, future plans, etc.

Active and Passive Sensors

Sensors are generally either active or passive. *Active* sensors transmit a signal that impinges on surrounding objects and the environment. An active sensor also receives and processes the reflection of this transmitted signal. Because of the time delay associated with two-way transmissions of information, active sensors can be used to determine the location of the target to some degree of accuracy. Examples of active sensors are Doppler radar, ground moving target indicator (GMTI) radar, and laser rangefinders.

Passive sensors act only as receivers and can detect a target only if it emits some sort of signal or if it emits radiant or reflected energy at a higher rate than does the background environment. Examples of passive sensors are signals intelligence receivers; electro-optical imaging devices, including traditional telescopes; and infrared sensors.

For many nonimaging passive sensors, a single sensor is not capable of determining the location of a target. Two or more nonimaging passive sensor target detections are needed to reduce target location error to a useful level. Thus, the fusion process is important for extracting useful surveillance and targeting information (and not just intelligence) from the output of many passive sensors. On the other hand, because they make use of the original emissions of the target or because they can produce high-resolution images of the target, passive sensors can greatly assist in identifying or classifying targets. High-resolution imaging sensors have traditionally been most valuable for locating and classifying targets. However, such sensors can be large and expensive and, if placed on airborne platforms, can be vulnerable to attack.

Sensor Detections

Sensors can typically detect targets or objects only when certain conditions prevail. For example, synthetic aperture radar (SAR) can detect images of stationary targets, but cannot detect moving objects. Target objects must be larger than some minimum size to be resolved in a SAR image, and target object features must be farther apart than the imaging resolution of the SAR imaging sensor for the target to be identified or classified. Similarly, a GMTI radar, when operating in this mode, can detect moving targets but cannot detect stationary targets. In addition, it can detect moving targets only if they are moving at a rate between some minimum and some maximum velocity. Infrared imaging sensors can detect a target, moving or stationary, only if its temperature is above some minimum relative to the temperature of the background environment.

Military operations typically use a wide variety of sensors. Each has its strengths and weaknesses. The generic model of sensor performance in this analysis is composed of probabilities of detection for sensors as a function of range. Target location and velocity errors (direction and speed) are also expressed as a function of range.

Completeness

The information gaps in the sensor subdomain relate to the number of enemy units detected, the features reported for the units detected, and the sensor suite's coverage area. For information to be complete, all features of all units in the relevant ground truth must be known, and the entire AO must be under sensor observation. This suggests the two measures and associated metrics.

The Detection of Units in the AO. It is assumed that the total number of enemy units in the AO is known. This is not too constraining given the IPB process, although this may vary by command echelon. Discriminating between real and false detections is a more serious problem. False detections can result from deception or from the inability of the sensor suite to distinguish between military and nonmilitary systems. GMTI radars, for example, will detect many objects that are not militarily relevant because they detect all ground vehicles moving above a certain speed. Reports from the current generation of GMTI sensors must be transmitted to fusion facilities (as discussed below) to distinguish military targets from other similar objects. The degree to which an individual sensor or suite of sensors can distinguish military targets from other similar objects is the subject of the correctness performance measure discussed below. What is relevant to the completeness measure is the degree to which military targets are *not* detected.

An appropriate metric corresponding to this measure is calculating or estimating the percentage of enemy units in the AO that have been detected. Developing an estimate requires a model of sensor performance and a model of the specific environment in which the sensor is operating. High-fidelity models of sensor performance typically include detailed descriptions of the environment in which the sensor operates and the targets the system is designed to detect and classify. High-fidelity models are applicable only for specific types of sensors. Thus, modeling the overall performance of a complete sensor suite at high fidelity requires a set of high-fidelity sensor models. Such an approach is beyond the scope of the present investigation and is not appropriate to the problem at hand. Appendix B describes a generic sensor model that can be calibrated to represent the performance of a wide array of possible sensors.

Sensor Suite Coverage. This simple calculation is based on the collective sensor areas of regard and their revisit rates. An appropriate metric, therefore, is simply the fraction of the AO covered. Redundant coverage, however, can cause the metric to assume values greater than 1.0. This is not a problem because this metric must be combined with the fraction of units detected to produce a composite completeness transfer function.

If we let c_1 and c_2 represent the metrics for the fraction of units detected and the fraction of area covered, respectively, combining the two metrics produces the completeness component of the transfer function $Q(F_1|F_0)$ depicted in Figure 3.1. Clearly, these two metrics are not independent. Complete or even redundant area coverage makes it more likely that a larger number of units will be detected and their features known. Redundant coverage also mitigates against false target detections. The following combining function reflects this:

$$Q_{com}\left(F_1 \mid F_0\right) = c_1\left(1 - e^{-c_2}\right).$$

With this formulation, the combined completeness score is always less than the fraction of units detected. However, as the coverage increases, it approaches but never exceeds that fraction. As the sensor area coverage increases, and becomes redundant, the fraction of real units detected more closely matches the total number detected. For example, if 70 percent of the units in the AO have been detected (real and false) and if the area coverage is 300 percent (that is, the sensor suite is capable of effectively sweeping the AO three times over), this combining function yields an estimate of the fraction of real targets detected of 0.665. We chose this form for the combining function because it both strongly penalizes completeness if not much of the AO is sensed (in which case, we might consider a high unit-detection score as being lucky, not good) and accurately indicates that there is little marginal benefit in increasing sensor coverage if most of the AO is already being "oversensed" (for example, increasing sensor coverage from 400 to 401 percent if of little benefit).

Correctness

The measures and metrics for correctness focus on deviations from ground truth. In an absolute sense, a report is completely correct if it matches ground truth exactly. For example, a report that an observed unit is mechanized is correct only if the observed unit is indeed mechanized. We recognize that, in some contexts, a report that the unit is armored would be a "correct" identification if the only concern were whether the unit has tracked vehicles. For our purposes, however, we have taken the more-absolute, context-free, approach and define a measure of correctness to be the degree to which the true target features approximate their ground truth values.

There are several ways to assess the nearness of an estimate to its ground truth value mathematically. We mention two here, one using *estimation* theory and the other assessing *precision*. The estimation theory approach is appropriate when analysts know ground truth directly (as in simulations or controlled experiments); the precision method is appropriate when analysts do not know ground truth (as in real-world operations).

Estimation. The entire sensor-collection process is essentially an attempt to *estimate* the ground-truth values of enemy unit features, intent, plans, and activities. It therefore seems natural to suggest that estimation theory might be useful for assessing the correctness of the estimate. If we view the collection of data about the enemy unit features as a collection of random samples, we can calculate an estimate of the true value of the feature using one of several statistical estimators. For example, suppose we wish to estimate the location of an enemy unit based on several sensor reports. Assuming that the location of the unit will be reported as an x-y coordinate, we need an estimate for both the x-coordinate and the y-coordinate. The simplest estimator is the sample mean. If $X_i = [x_i, y_i]$ represents the ith sensor report estimate for the enemy unit, the following is a simple estimate of its true location:

$$\hat{\mu} = \frac{1}{n} \sum_{i=1}^{n} X_i \, ,$$

where n is the total number of reports arriving.

The chore now is to assess just how correct this estimate is. If we were conducting a controlled experiment, such that the true location of the enemy unit is known, then we could take advantage of the fact that an unbiased estimator is one in which

$$E\left[\hat{\mu}\right] = \mu \, ,$$

where $\mu = [\mu_x, \mu_y]$ is the true, known, mean value. For our purposes, we can take this to be the true location of the unit. The correctness of the estimate in each direction can be calculated to be

$$A = \left| E\left[\hat{\mu}\right] - \mu \right| \, ,$$

generally referred to as the *bias* in the estimate.[1] Calculating

$$E\left[\hat{\mu}\right]$$

---

[1]Note that this refers to statistical bias, not operational bias. We assume that the estimates of features have been corrected for sensor bias.

requires several samples of sensor reports or the partitioning of a single sample. For nonquantitative features, the mode of all reports might be used as the estimator. However, calculating the bias is problematic. One technique is to develop similarity matrices.[2] One way to use the bias to calculate the correctness component of the transformation function in Figure 3.1 is

$$Q_{cor}\left(F_1 \mid F_0\right) = e^{-\left(A_x + A_y\right)},$$

where A_x and A_y are the biases in the x- and y-position estimates, respectively. If we take the estimate to be completely correct when $Q_{cor}(F_1|F_0) = 1$, correctness gets worse for large biases, as expected. The use of the exponential form rapidly penalizes correctness as bias errors initially increase; correctness levels out as the biases become large enough to impair mission effectiveness, such that any additional bias has little effect.

Several estimators other than the sample mean might be used. One of these, the linear estimator, is a linear function of the sensor observations. For example, we can estimate the location of a moving enemy unit at some time in the future by observing its current location and its velocity. Multiplying the x- and y-components of the velocity by the time elapsed added to the current estimate makes the location of the enemy unit a linear function of its velocity. Maximum likelihood and minimum mean-square estimators can also be used, as appropriate.[3]

Precision. Bias appears to be a good way to measure correctness when conducting an experiment in which the true value of the feature to be estimated is known. However, assessing the correctness of sensor observations when the true values of the features are not known (as in support of combat operations) requires using other methods. One possibility is to calculate the precision of the reports. *Precision* is the ability of a sensor suite to provide repeated estimates that are very close together. This is another way of saying that some

---

[2]Appendix B offers a methodology for producing such a matrix.

[3]Any standard statistics text can provide more details on these estimators. See, for example, Stark and Woods (1986).

function of the variance in the sample of estimates might serve as a metric. Continuing the above example, suppose now that the ground truth location of the enemy units is not known and that the sensors reported 27 locations for enemy units, as depicted in Figure 3.2. Using a suitable cluster algorithm indicates that the reports are widespread enough to suggest that there are four enemy units in the observed area.[4] The algorithm produces a *cluster representative* location, depicted as

$$\overline{\mathbf{X}}_i$$

in the diagram. The cluster representatives can then be taken to be the location estimates for the enemy units.

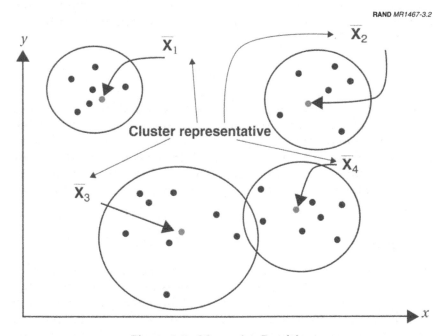

RAND MR1467-3.2

Figure 3.2—Measuring Precision

_____

[4]Clustering is used when no prior information is available on the disposition of the enemy forces. With prior information, the observations "cluster" around that previous information. For a complete discussion of cluster algorithms, see Gordon (1999) and Duda and Hart (1973).

The location of each enemy unit is taken to be a bivariate normal random variable whose mean is the ground truth location of the unit and whose covariance is estimated by the following sample covariance matrix:

$$\hat{\Sigma} = \begin{bmatrix} S^2 & 0 \\ 0 & S^2 \end{bmatrix} = \frac{1}{n-1} \Sigma_{j=1}^{n} \left[\mathbf{X}_j - \bar{\mathbf{X}} \right] \left[\mathbf{X}_j - \bar{\mathbf{X}} \right]^T .$$

In this formulation,

$$\bar{\mathbf{X}} = \left[\bar{x}, \bar{y} \right],$$

$\mathbf{X}_j = [x_j y_j]$ represents the other cluster locations, and S^2 is the sample variance in both the x- and y-directions. The zeroes in the off-diagonal positions reflect the fact that the x- and y-positions of the enemy unit are independent. Precision is then defined to be the determinant of the covariance matrix:

$$p = \left| \hat{\Sigma} \right| = S^4 .$$

Note that this value is always nonnegative and that a 0 value implies perfect precision. Therefore, the correctness component of the transformation function in Figure 3.1 might be $Q_{cor}(F_1|F_0) = e^{-p}$. If we continue the same convention as with the bias metric—that complete correctness occurs when $Q_{cor}(F_1|F_0) = 1$—correctness gets worse for large values of p. Since precision is exact when $p = 0$, this produces the desired effect.

Currency

In general, we assumed that the less time required to complete a process the better. Therefore, currency measures generally assess the time required to perform functions. For the sensor subdomain, the measure is the time required to complete sensor operations and local data processing. The number of tasks this requires clearly depends on the situation. Some may be accomplished in parallel, while others might be sequential. In any case, what is required is an estimate of the total time elapsed between detection and receipt of the report at the fusion center.

Suppose two tasks consume time for a given sensor suite:

- completing target detection and establishing a target track for an enemy unit
- retasking sensors to provide additional coverage of units to reduce uncertainty in the estimates.

The total time expended in getting a set of reports to the fusion center is dictated by the report that takes the longest. The methodology used to assess this time is referred to as the critical path method. Because this same method is used for all currency calculations, details are included in Appendix B.

Suppose that the time required to get a report containing several observations based on initial detections and retaskings to the fusion facility is t. In the absence of any other situation-dependent information, we assume that currency is greater for smaller values of t, so that a task that is accomplished instantaneously has maximum currency. Applying logic similar to that applied to the completeness and correctness metrics yields a currency component of the transformation function in Figure 3.1 of $Q_{cur}(F_1|F_0) = e^{-t}$. In addition to mapping the reporting time to a 0–1 value as desired, with 1 being best, using the exponential function heavily emphasizes completing the report quickly while not greatly distinguishing between lengthy reporting times.

THE FUSION SUBDOMAIN

The output of the sensor subdomain is a series of sensor reports that are forwarded to fusion centers for processing. For purposes of this study, we assumed that reports from like and disparate sensors and sources would be combined to produce a CROP, which would subsequently be disseminated to friendly users. The quality of the information the process produces is represented by $Q(F_0,F_1,F_2)$ in Figure 3.1. The correctness and completeness of the process are conditioned on the quality of the information and data received from the sensor suites, represented by $Q(F_0,F_1)$ in Figure 3.1.

Fusion

Fusion is the process of combining information from sensors and sources to produce a common, relevant picture of the battlespace. It includes the correlation and analysis of data inputs from supporting sensors and sources. The relevant picture drawn can vary depending on how the fused information is to be used. If, for example, the information is to be used only for targeting, simply combining reports to pinpoint the location of enemy targets is sufficient. If the information is to be used to develop plans for maneuver, the reports must include some assessment of enemy intent, plans, and activities. Either use ultimately leads to a decision. In the first case, targeting decisions are generally made at the tactical level; in the second, maneuver decisions are generally made at the strategic and operational levels of combat.

This suggests a fusion taxonomy that focuses on the degree to which the fusion process produces inferences. Table 3.2 lists five *levels* of fusion that constitute a spectrum of fusion. The lower levels (0 and 1) represent what is generally included in the CROP today. It consists mainly of individual target locations and refined tracks posted on an area map. No attempt is made to infer unit affiliation, enemy posture or intent. This level of fusion is generally sufficient to support target acquisition, but its contribution to situation assessment is minimal. At the higher levels of fusion, fewer burdens are placed on the decisionmaker in that he is not as required to draw inferences about capabilities and intent. Automation at these levels is rare and therefore when it is done at all, it is usually a manual process conducted by the unit intelligence element.

Table 3.2

Fusion Processing Levels

Processing Level	Description
0	Normalizing, formatting, ordering, and compressing input data
1	Refining position, tracks, and identity estimates
2	Interpreting relationships between objects and events; developing situation estimates
3	Assessing enemy capability and intent
4	Continuously refining estimates; identifying needs for additional sources and processing

SOURCE: These definitions were adapted from Keithley (2000).

As mentioned earlier, however, this study is restricted to those aspects of ground truth that focus on the relevant features or characteristics of enemy units and not on enemy intent, plans, and activities. In that sense, therefore, the fusion we are dealing with is at levels 0 and 1.

Uses of Fused Information

The lower levels of fusion are generally characterized by automated systems, such as the Army's Q-37 Firefinder radar system. The Q-37 detects enemy artillery and mortar fires and relays the information to the fire control center within seconds. Counterbattery fires are then launched at the enemy weapon systems. Sensors that detect mobile missile launchers pass their information to orbiting combat air patrol aircraft, which then attempt to destroy the launchers. The level of fusion in these cases is rather low and is generally completed by the sensor system itself. For example, an MTI combines location, speed, and direction of movement to help establish a track for the observed target. This might apply to tracking the moving missile launcher.

Assessing the enemy situation at the higher levels of fusion is more complex, generally consisting of both automated and manual processes. Much more information may be needed than that required for target nomination. For example, it may not be enough to know where the enemy is; it may also be necessary to know what he intends to do next or what he is capable of doing next. These assessments generally take place in the command operating center as an adjunct to the intelligence process. However, both distributed and centralized systems are possible. To provide input to what is in essence a manual process, the Army has developed and implemented the All Source Analysis System, a centralized fusion center where data from all sources are combined, correlated, analyzed, and disseminated to eligible users. Other systems have also provided input to the process. In the mid-1980s, U.S. Air Forces Europe initiated the Enemy Situation Correlation Element to do the same thing. At a lower level, the Joint Surveillance Target Advance Radar System (STARS) Ground Station will include the Joint STARS Work Station, an automated MTI fusion system. None of these systems in itself provides higher levels of fusion. This is generally performed by the command intelligence staff.

Fusion Facilities

Fusion is essentially a parallel-sequential process. Each intelligence discipline (INT) attempts to fuse its internal estimates and forward them to a central fusion processor, which combines fused feature vectors from disparate INTs to develop the CROP. Although this may not be the physical model in all cases, the basic sequence is generally applicable. Figure 3.3 illustrates the process.

It is generally reasonable to assume that the longer a fusion facility has to examine the sensor or within-discipline reports, the more reliable the results will be. This assumption is based on the possibility that the longer the fusion facility has to complete its processing, the more information will be available. Also, the results of some time-intensive activities, such as image processing, will improve given more processing time. This also accounts for the time required to retask or cue sensors to focus on targets of interest.

Automation and Control

The rate at which the center classifies detections depends in part on the degree of automation at the fusion facility. The minimum frac-

RAND MR1467-3.3

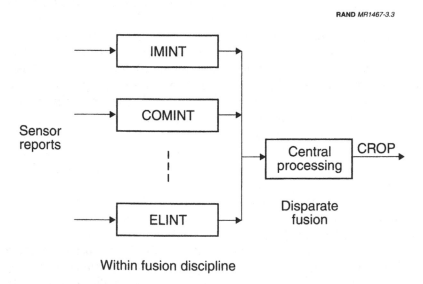

Figure 3.3—Parallel-Sequential Fusion Process

tion of detections that can be classified depends on the characteristics of the sensor suite, but the maximum depends on the ability of the fusion center to control the supporting sensor suite. We treat automation as a binary variable: Either the facility is automated or it is not. We distinguish two types of operational control: dynamic and static.

Static Control. If the fusion facility can only control the sensors by providing their initial tasking, it has *static control* over the sensors. A sensor management plan is established at the outset of operations, and it essentially remains fixed. This has the effect of slowing the classification process if the plan was deficient in any way. Therefore, the fraction of classified detections rises slowly over time.

Dynamic Control. If the fusion facility can task and retask the sensors to confirm reports or to bridge gaps in the data, it has *dynamic control* over the sensors. As with static control, the fusion facility begins with a sensor management plan but, in this case, can alter the plan as the operation progresses.

Completeness

The ultimate goal of the fusion process is to create a picture of the battlefield that is complete, accurate, and current. The "picture" consists of sets of estimated values for the features of enemy units. The objective of the process is to *classify* the unit estimates by describing the units in terms of the relevant features. The completeness portion of the quality measure, then, focuses on the number of these units contained in the CROP.

We recognize that some degree of classification can take place at the sensor level. For example, cueing an unmanned aerial vehicle to fly over a location to view a detection made by Joint STARS is an attempt to classify the detected object using only a sensor. Assuming this has been accomplished for one or more of the targets, the chore at the fusion facility is easier, in that the ability to "classify" has been enhanced. Although classification can take place at both the sensor and the fusion facilities, we chose to evaluate its quality at the fusion center to clearly delineate functions.

Classification Rate. The rate at which a fusion facility can classify detected units greatly influences the number of detections the facility can classify in a fixed period. This depends on several factors that are certainly situation dependent. However, it may be possible to resolve all these factors into a simple expression, $c = f(t)$. That is, the fraction of detections that can be classified, c, is dependent on the time available, t. One such representation is an increasing exponential. Each of the within-discipline fusion centers and the central fusion center are assigned a time-dependent fraction of detections classified of the form

$$c = a + \alpha\left(1 - e^{-\sigma t}\right),$$

where a represents the fraction of the detected targets that the sensors themselves can fuse; $a + \alpha$, for $0 < a + \alpha \leq 1$, is the maximum fraction of the detections capable of being classified at the fusion center; and σ is the rate at which detections are classified. Verifying that the functional relationship is appropriate and obtaining realistic values for the parameters can be formidable.

To illustrate, Figure 3.4 provides several examples. For each curve, a and α are fixed at 0.2 and 0.7, respectively, and only the rate of classification, σ, changes. Note that, in this example, the fraction of classified detections for all centers is bounded between $a = 0.2$ and $a + \alpha = 0.9$.

Automation and Control Effects. Combinations of automation and control protocols can affect the classification rate at each of the fusion facilities. The effects are therefore reflected in the values of the parameters a, α, and σ. Assuming that the functional relationship between the fraction of detections classified and the time is correct, it follows that automation and control will greatly affect the shape of the curves depicted in Figure 3.4.

The overall completeness of the fusion process for k INTs can therefore serve as a reliability model. The facilities depicted in Figure 3.3 process estimates in parallel and feed their results to the central processor in series. If we assume (1) that the time available to process reports at each of the fusion facilities is fixed, (2) that the detections each fusion facility processes are randomly and independently selected, and (3) that a report on a detection from a single fusion facility is necessary and sufficient to classify that detection at the central processing facility, the following provides a measure of completeness for the simple configuration depicted in Figure 3.3:

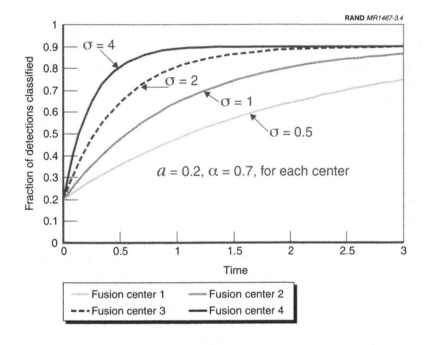

Figure 3.4—Fusion Center Classification Rates

$$C = \left[1 - \prod_{i=1}^{k}(1-c_i)\right]c_c \, ,$$

where c_i is the fraction of the detections that fusion facility i can process in the time available, and c_c is the same quantity for the central
processing facility. This quantity, then, is taken to be the completeness component of the transformation function depicted in Figure
3.1: $Q_{com}(\mathbf{F}_2|\mathbf{F}_0,\mathbf{F}_1) = C$.

Correctness

As in the sensor subdomain, the appropriate measure of correctness
is how close the fused estimate for each enemy unit feature is to
ground truth—how accurate the classifications of the detections
reported are. As before, the problem is assessing how good the estimate, and therefore the fusion process, is. However, in this subdomain, we added the additional task of tracking enemy units. This
suggests two measures: (1) how well the fused features of the units

reflect ground truth, i.e., how well detections can be classified, and (2) how well the fusion system tracks enemy units over time.

Classification. Sensor reports arriving at fusion facilities may or may not confirm previous reports. Confirming evidence tends to reduce uncertainty, and disconfirming evidence tends to increase it. In either case, the variation among the reports appears to be central to measuring how well the fusion process can reflect ground truth. With minimal disconfirming reports arriving concerning the features of the detected units, the variance in the estimates will likely be small, therefore increasing the likelihood that a "correct" classification will result from the fusion process. The opposite is true if the number of disconfirming reports is large.

This suggests that one way to address this problem might be to examine the variance in the feature estimates for each reported unit. For example, suppose the fusion process is to classify enemy units based on three reported features: location, velocity, and unit type. Therefore, in this case, correctness measures how close estimates of these three features are to ground truth for each enemy unit included in the reports. Further suppose that, for a given unit, there are 10 reports containing estimates for all three features. Location and velocity are bivariate (x- and y-position for location, and speed and direction for velocity); therefore, the determinant of the sample covariance matrices, $|\Sigma_l|$ and $|\Sigma_v|$ for location and velocity, respectively, would be a measure of the variance among the estimates. This is similar to our previous assessment of precision for sensor reports. The estimates of unit type are a bit more problematic, in that the types are nominal. Nevertheless, a similarity matrix can provide an estimate of the variance, as described in Appendix B.

This approach still requires two more levels of combination: a single measure of overall variance for each enemy unit reported and a measure for the variance among all the reported units. Both require combining disparate information. Fortunately, the sample variance is always positive. Therefore, if we create the metric $a_i = e^{-s_i}$, where s_i is the sample standard deviation for the ith feature of the unit, a_i is close to 1 for small variances (better correctness). For large variances, a_i approaches 0. A simple weighting scheme might then lead to a unit composite score. In our simple example, this might yield

$$W_i = \sum_{j=1}^{3} \omega_j a_j ,$$

where

$$\sum_{j=1}^{3} \omega_j = 1.$$

This produces a quantity between 0 and 1, where a value close to 1 is desirable. The problem is the basis for selecting the weights. One criterion might be the relative importance of the feature in targeting the enemy unit.

We might do something similar for the next combining level, assessing the effects of report variances among all the enemy units in the reports. Establishing the weights here may also be a bit problematic. This requires judging the relative importance of classifying the various enemy units. Nevertheless, if it can be done, we would again have a metric,

$$W = \sum_{i=1}^{n} w_i W_i \, ,$$

where

$$\sum_{j=1}^{n} w_i = 1.$$

This also produces a quantity between 0 and 1. This quantity then can be taken to be the classification portion of the correctness component of the transformation function depicted in Figure 3.1: $Q_{cor,1}(\mathbf{F}_2|\mathbf{F}_0,\mathbf{F}_1) = W$.

Tracking. Much has been written about tracking, notably by Samuel Blackman in a survey of the subject he published in 1986. Blackman was interested in multiple-target tracking, which is applicable to tracking the movement of enemy units in the battlespace. Tracking is recursive in that observations or detections of enemy units must be correlated over time. The basic elements of a recursive tracking system are depicted in Figure 3.5.

For this example, we assumed that information is available from a previous collection cycle, referred to as a *scan*. First, the input is examined to see if it correlates with existing tracks. The gating function is a coarse assessment to determine whether the detection is a candidate for a track update or is the basis for a tentative new track. If the former, a more refined detection-to-track pairing algorithm is

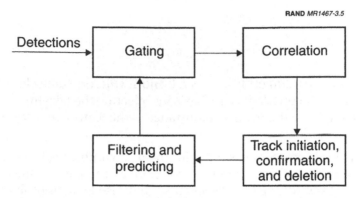

SOURCE: Adapted from Blackman (1986) p. 5.

Figure 3.5—Basic Elements of a Recursive Tracking System

used to make the final pairing. A tentative track is confirmed if the number and quality of the detections meet established criteria. A track that is not updated becomes degraded and must be deleted. Filtering incorporates the correlating detections to create an updated estimate. Prediction extrapolates the track parameters to the next scan. If no update is available, the previous update is used.

The correctness of the tracks of enemy units that this or a similar process produces can be measured in terms of the number of previous tracks that have been confirmed during this scan and the number of new tracks initiated. For simplicity, we assumed that tracks are never deleted but, rather, are placed in an inactive status.

If tracks exist for k enemy units current to scan $t-1$ and n exist at the current scan, t, we must examine two cases: $n = k$ and $n \geq k$. A third, the deleted-tracks case ($n < k$), would not apply given our assumption that deleted tracks are simply put in an inactive status. If we let X_{t-1} represent the set of tracked enemy units at scan $t-1$ and X_t the set of tracked enemy units at scan t, we have the following two cases:

a. If $n = k$, we either are able to correlate the tracks in X_{t-1} with the tracks in X_t or are not able to do so. Three possibilities arise, as depicted in Figure 3.6.

b. If $n > k$, there are more tracks at scan t than at scan $t-1$. This suggests the three cases depicted in Figure 3.7.

These two cases allow us to specify the fraction of the enemy units tracked at scan $t-1$ that are correlated with detections at scan t as a metric to measure how well the system is able to track enemy units over time. In Figures 3.6 and 3.7, $T = 1.0$ for A and D; $T = 0$ for B and E; and $0 < T < 1$ for C and F.[5]

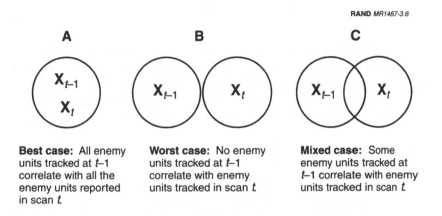

A   **B**   **C**

Best case: All enemy units tracked at t–1 correlate with all the enemy units reported in scan t

Worst case: No enemy units tracked at t–1 correlate with enemy units tracked in scan t

Mixed case: Some enemy units tracked at t–1 correlate with enemy units tracked in scan t

Figure 3.6—Tracking Cases When $n = k$

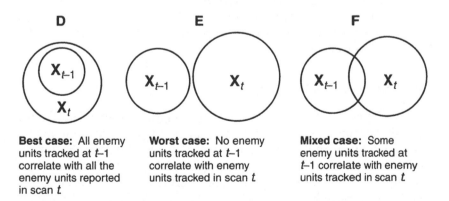

D   **E**   **F**

Best case: All enemy units tracked at t–1 correlate with all the enemy units reported in scan t

Worst case: No enemy units tracked at t–1 correlate with enemy units tracked in scan t

Mixed case: Some enemy units tracked at t–1 correlate with enemy units tracked in scan t

Figure 3.7—Tracking Cases When $n > k$

---

[5]Appendix B includes a set theoretic discussion of this metric.

This quantity, then, can be taken to be the tracking portion of the correctness component of the transformation function depicted in Figure 3.1: $Q_{cor,2}(\mathbf{F}_2|\mathbf{F}_0,\mathbf{F}_1) = T$. We might combine the two using an importance weight, $0 \le \omega \le 1$, so that the correctness component of the transformation function is $Q_{cor}(\mathbf{F}_2|\mathbf{F}_0,\mathbf{F}_1) = \omega W + (1 - \omega)T$.

Currency

Currency can be defined as the total time required to update the CROP using inputs from the integrated multisensor suite. If t is the total time required to update the CROP, $Q_{cur}(\mathbf{F}_2|\mathbf{F}_0,\mathbf{F}_1) = e^{-t}$ provides a metric for the currency component of the transformation function depicted in Figure 3.1. Instantaneous updating earns a score of 1.0.

THE NETWORK SUBDOMAIN

There are two networks to contend with: the communications network that supports the multisensor system and the network that supports the dissemination of the CROP to required users. The latter is the network subdomain we discuss here, through which the CROP developed in the fusion subdomain, $\mathbf{F}_2$, is disseminated via a communications network that connects all users to the fusion facilities. This product, the observed CROP, informs both the commander's situational awareness and the decisions he is to take. The quality of the observed CROP is conditioned on the information transmitted from the fusion facilities.

Communications Networks

The battlefield of the future is likely to be highly dispersed, and combat will therefore be nonlinear. This places considerable demands on communications networks that support C⁴ISR functions. For example, is it more efficient to create a single, perhaps out-of-area, fusion center, or are distributed centers more efficient? If there is a robust reachback capability, it can be argued that considerable efficiencies are possible if fusion resources are concentrated at a single site. But this takes the responsibility for developing the CROP out of the hands of the local commander and can easily foster a "not invented here" attitude. Indeed, this occurred during the Kosovo conflict. The Commander of Task Force Hawk rejected the opinion of the Joint

Assessment Center at Molesworth, England, that the threat from the 2nd Yugoslav Army in Montenegro was minimal in favor of his own that this army posed a threat to his Apache helicopters based at Rinas, Albania.[6]

On the other hand, a distributed system also has its problems. The widely varying and incomplete resources at some of the sites clearly might mean that the quality of the fusion process could be uneven. For example, it is likely that not enough imagery analysts will be available at enough sites within a theater to support local commanders adequately. In addition, it is likely that there will not be enough bandwidth available to allow all sites to support local fusion. The demands on the communications networks in either case are considerable.

Completeness

The measure of completeness for the network subdomain is essentially a network reliability measure applied to the dissemination of the CROP to all its users. Not all users require the complete CROP, and not all require the same level of resolution. However, the users must all share the same information. *Completeness* therefore measures how well the communications network accommodates the transmission of relevant aspects of the CROP to each user. It addresses the likelihood that each user will receive the relevant portions of the CROP.[7] A possible metric for this measure is the probability that all users will receive the CROP. This is an assessment of the network's reliability in terms of its robustness.

Several factors affect the reliability of a communications network, foremost among them the degree to which it provides alternative transmission routes. Other factors include the degree to which facilities are hardened, enemy jamming and other electronic warfare attacks, and bandwidth. To develop a metric for overall network reliability, it is necessary to (1) assess the probability that each user can

---

[6]This and other incidents concerning the deployment of Task Force Hawk during Operation Allied Force is documented in Nardulli et al. (2002).

[7]The subsection below on correctness discusses the likelihood that it will be "correctly" received and that the same information is received.

receive a message, then (2) combine these to assess the probability that all users can receive a message. The first step requires the assessment of link reliabilities for all paths from sender to receiver. The second implies the assessment of a joint probability.

Suppose we let p_i represent the probability that user i will receive the CROP from the central fusion facility, based on all the possible routes it could take. The overall network reliability therefore would be

$$N = \prod_{i=1}^{k} p_i \, ,$$

where k is the total number of users. Implicit in this product is an independence assumption: that the probability that user i receives the message is independent of the probability that user j receives it.[8]

The completeness component of the transformation function depicted in Figure 3.1 would therefore be $Q_{com}(\mathbf{F}_3|\mathbf{F}_0,\mathbf{F}_1,\mathbf{F}_2) = N$.

Correctness

Network correctness can be measured in terms of the likelihood that all users receive the same CROP or portion of the CROP that was transmitted to each from the central fusion facility. This is an assessment of the likelihood that CROP users receive the distributed information without degradation.

One way to measure this likelihood is to use the probability of correct message receipt (PCMR) as a metric. As with completeness, using PCMR as a metric requires a two-step assessment. First, we determine the PCMR for each user, then expand it to include all users. The PCMR is essentially a conditional probability that the message sent will be the message received. Applied to the dissemination of the CROP, the probability that user i will receive the CROP (or portion of it) as transmitted from the central fusion facility is $P_i(\mathbf{F}_3|\mathbf{F}_2)$. As with the completeness metric, these probabilities are related to the quality of the channel(s) over which the information is transmitted. Therefore, such things as signal-to-noise ratio (SNR), bandwidth, and jamming are also determinants of the PCMR. However, the value is

---

[8]Appendix B develops the reliability estimates for the links in more detail.

also dependent on the input to the channels. The PCMR, then, is the joint probability that the central fusion facility transmitted the CROP correctly and that the user received the observed CROP correctly: $P_i(F_3,F_2) = P(F_2)P_i(F_3|F_2)$.[9]

The problem, then, is to find adequate representations for the marginal probability, $P(F_2)$, and the conditional probabilities, $P_i(F_3|F_2)$. The second of these is taken to be dependent on the reliability of the communications paths between the central fusion facility and the user. This is the connectivity probability, p_i, discussed above, under completeness. The marginal probability, $P(F_2)$, is the probability that the CROP, or portion of it, will be transmitted correctly and is therefore a function of the communications equipment and personnel within the fusion facility. Given that these probabilities can be assessed, the PCMR for user i is $PCMR_i = P_i(F_3,F_2) = P(F_2)p_i$.

The overall network PCMR can be thought of as the joint probability of all the fusion facility-to-user PCMRs:

$$PCMR = \prod_{i=1}^{k} P(F_2)p_i .$$

The correctness component of the transformation function depicted in Figure 3.1 would therefore be $Q_{cor}(F_3|F_0,F_1,F_2) = PCMR$.

Currency

Network currency depends on the rate at which data can be transmitted over the network's links and the amount of time required to process information at the various nodes in the network. The former depends on bandwidth and the complexity of the communications paths between subscribers, and the latter depends on the activities performed at the nodes. The end-to-end time required to transmit the CROP from the central fusion facility to the users can therefore serve as a measure of network currency.

---

[9]See, for example, Blahut (1988) for further discussions of probability distributions for sources and channels.

Calculating these times can be formidable. Depending on the degree of connectivity in a network, the number of ways to get a message between two nodes can be considerable. Given the myriad ways to get a message between two nodes, how should we assess the end-to-end transmission time? There are several possibilities. A conservative approach is to calculate the time required along the "longest" path. Another is to select the minimum, and still another might be to calculate the average of all paths. For this analysis, we chose the average transmission and processing times. The overall average network transmission delay, then, is the average of these times, $\bar{t}$, so that the currency component of the transformation function in Figure 3.1 is

$$Q_{cur}\left(\mathbf{F}_3 \mid \mathbf{F}_0, \mathbf{F}_1, \mathbf{F}_2\right) = e^{-\bar{t}}.$$

An Alternative for Complete Networks

The methodology we have suggested for calculating completeness, correctness, and currency of a communications network becomes rather tedious when the size and connectivity of the network increase. In the limit, for example, every node is a relay, and every entity on the battlefield is a node. The Army refers to this as a "nodeless network," the idea being that, with such a richly connected network, the loss of a node will not degrade performance in any appreciable way.[10]

The fully connected, or *complete*, network is a subset of the nodeless network. A complete network of n nodes, in which all connections are two-way, has $n(n-1) = n^2 - n$ connections. Since we are interested only in alternative paths from one node to another, a large n is desirable. Consequently, *given that we have a complete network*, a surrogate for network quality in all of its dimensions might be the size of the network. For a very large n, the first term, n^2, dominates, and the quality of the network therefore increases as the square of the number of nodes in the network. This is referred to as *Metcalf's Law*,

---

[10]Taken from a briefing presented by the U.S. Army Communications-Electronics Command Research Development and Engineering Center, Fort Monmouth, New Jersey (Nichols, 1999).

named for Robert Metcalf, a pioneer in the development of the Ethernet.[11] The transformation function for currency can then be written as

$$Q_{cur}\left(\mathbf{F}_3 \mid \mathbf{F}_0, \mathbf{F}_1, \mathbf{F}_2\right) = 1 - e^{-n^2} .$$

In addition to mapping the size of the network to a 0–1 value, the exponential form of the function rapidly approaches 1 initially, then levels off as the size of the network increases—a reasonable assumption, since, after a certain size, node losses are unlikely to degrade the network significantly.

Shared Information

The CROP produced at the central fusion facility is distributed to the users over the dissemination network; that is, it is shared among the users. Shared information is essential for effective collaboration, as we will discuss more fully in Chapter Four. However, because effectively sharing information depends so much on the structure of communications networks, it seems appropriate to elaborate a bit here.

Earlier, we referred to the CROP the users receive as the *observed* CROP. Recall that matrix $\mathbf{F}_2$ represents the *fused* CROP, which contains estimates of the features or characteristics of the enemy units detected. Not all friendly users need all the information contained in the perceived CROP. So, each user's observed CROP is a subset of the fused CROP.

Figure 3.8 depicts the amount of information the users receive. The shaded circles in the diagram represent the amount of information (the portion of the CROP) that each of three users receives. The values on the edges represent the probability that each user will receive the relevant portions of the CROP. The white circles represent the entire CROP. The small dark gray area in the center of the three

---

[11]Metcalf also founded the 3Com Corporation of Santa Clara, California, in 1981, the leading producer of Ethernet adapter cards. Gilder (1993) provides an interesting discussion of Metcalf and his "law of the telecosm." See also Appendix A to Alberts, Garstka, and Stein (2000).

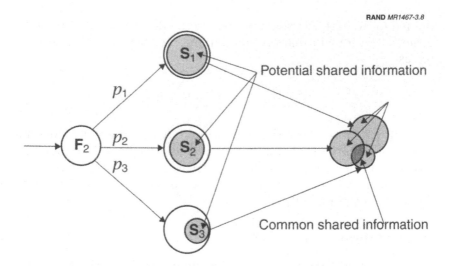

Figure 3.8—Shared and Common Information

joined circles represents the information that is common among all the users, while the residual gray area has the potential to be shared among all the users, depending on the ability of the group to collaborate. It is clear that the ability to collaborate has the potential to increase the amount of information shared among the users thus contributing to shared situational awareness, as discussed in Chapter Four.

SUMMING UP

In this chapter, we applied the methodology we developed in Chapter Two to the physical and information domains. We introduced the feature matrix, $\mathbf{F} = [F_1, F_2, \ldots, F_m]$, as a set of vectors, F_i, each of which represents the relevant features of the enemy. In the physical domain, $\mathbf{F}_0$ represents the ground truth features of all enemy units. The information domain has three subdomains: sensor, fusion, and network. This chapter assessed the three components of information quality—completeness, correctness, and currency—for the information product generated in each subdomain.

The quantities developed here, in subsequent chapters, and in Appendix B are illustrative and do not represent settled research.

Sensor Metrics

In the sensor subdomain, we examined three aspects of completeness: the number of enemy units detected, the features reported for the units detected, and the sensor suite's coverage area. For information to be complete, all features of all units in the relevant ground truth must be known, and the entire AO must be under sensor observation. The suggested completeness metric consists of two components, both of which are between 0 and 1: c_1 is the fraction of enemy units detected, and c_2 is the fraction of the AO covered. The proposed completeness transformation metric then is

$$Q_{com}(\mathbf{F}_1|\mathbf{F}_0) = c_1(1-e^{-c_2}).$$

We suggested two types of metrics for correctness, those designed to support controlled experiments and those designed to support operations, and defined *correctness* as the degree to which the true target features approximate their ground truth values. We suggested using estimation theory to assess the deviation from ground truth for controlled experiments. Since an unbiased estimator of a parameter is one whose expected value matches the true parameter, the difference between the estimate and the known ground truth appears to be suitable for measuring correctness. In general, if A is a measure of nearness, such that near-zero values of A correspond to high nearness, the proposed correctness transformation is $Q_{cor}(\mathbf{F}_1|\mathbf{F}_0) = e^{-A}$.

Assessing correctness in support of operations implies that ground truth is not known. In this case, we clustered the detections geographically using a pattern-classification technique and then calculated the variance within the cluster. For a location estimate, this variance is expressed in terms of a covariance matrix. The determinant of that matrix is a measure of precision and, therefore, a measure of correctness. The determinant is $p = S^4$, where S^2 is the sample variance in both the x and y directions. The proposed correctness transformation then is $Q_{cor}(\mathbf{F}_1|\mathbf{F}_0) = e^{-p}$.

Fusion Metrics

Fusion is the process of combining information from sensors and sources to produce a common, relevant picture of the battlespace, the CROP. This process includes the correlation and analysis of data inputs from supporting sensors and sources. Fusion occurs at several levels, from the simple combining of tracks and identity estimates to assessments of enemy intent. Our focus here is on the lower levels of fusion.

We assumed that the architecture of the fusion suite consists of several fusion facilities that focus on a single intelligence discipline. These facilities transmit their fused reports to a central facility that produces a final product, the CROP.

Completeness in the fusion subdomain focuses on the number of detected enemy units from the sensor subdomain that are classified, i.e., described in terms of the relevant features. The number the fusion facilities can classify depends on the architecture of the fusion suite, the degree of automation, and the control procedures (the ability of the system to retask the sensors). The proposed completeness transformation is

$$Q_{com}\left(\mathbf{F}_2 \mid \mathbf{F}_0, \mathbf{F}_1\right) = \left[1 - \prod_{i=1}^{k}\left(1 - c_i\right)\right] c_c ,$$

where c_i is the fraction of the detected enemy units fusion facility i can process per unit time and c_c is the fraction the central processing facility can process.

Correctness in the fusion subdomain measures how close the fused estimate for each enemy unit feature is to ground truth—that is, how accurate the classifications of the reported detections are. One way to address this problem is to examine the variance in the feature estimates for each reported unit. This results in the following correctness transformation:

$$Q_{cor,1}\left(\mathbf{F}_2 \mid \mathbf{F}_0, \mathbf{F}_1\right) = \sum_{i=1}^{n} w_i \sum_{j=1}^{p} \omega_j \left(e^{-s_j}\right) .$$

where w_i and ω_j are weights, the first accounting for the relative importance of the reported enemy unit and the second for the relevant importance of the features being reported; s_j represents the sample standard deviations for each of the p features for a given enemy unit, derived from the number of reports arriving on the unit;

and the second subscript on Q distinguishes this correctness transformation from the tracking metric (discussed next).

We also chose to measure how well we are able to track enemy units. The correctness of the tracks of enemy units can be measured in terms of the number of previous tracks that have been confirmed on the present scan and the number of new tracks initiated. We took the tracking portion of the correctness component of the transformation function to be $Q_{cor,2}(F_2|F_0,F_1) = T$, where T is the fraction of the enemy units that correlate with previous tracks.

Combining the two correctness metrics using an importance weight, $0 \le \omega \le 1$, we get the following correctness component for the transformation function: $Q_{cor}(F_2|F_0,F_1) = \omega W + (1 - \omega)T$.

The metric for the currency component of the transformation function might be $Q_{cur}(F_2|F_0,F_1) = e^{-t}$, where t is the total time required to update the CROP.

Network Metrics

Completeness measures how well the communications network accommodates the transmission of relevant aspects of the CROP to each user. The metric we chose for this measure is the probability that all users will receive the CROP. This is an assessment of the network's reliability in terms of its robustness. The completeness component of the transformation function is

$$Q_{com}\left(F_3 \mid F_0,F_1,F_2\right) = \prod_{i=1}^{k} p_i \, ,$$

where k is the number of users of the CROP, and p_i represents the probability that user i will receive the CROP.

Network correctness is an assessment of the likelihood that CROP users will receive the distributed information without degradation. One way to measure this is to use the PCMR. The PCMR is a conditional probability that the message sent will be the message received. The probability that user i will receive the CROP (or portion of it) as transmitted is $P_i(F_3,F_2) = P(F_2)P_i(F_3|F_2)$, where $P(F_3|F_2) = p_i$. Therefore, the PCMR for user i is

$$\text{PCMR}_i = P_i\left(\mathbf{F}_3, \mathbf{F}_2\right) = P\left(\mathbf{F}_2\right) p_i$$

and the correctness component of the transformation function is

$$Q_{cor}\left(\mathbf{F}_3 \mid \mathbf{F}_0, \mathbf{F}_1, \mathbf{F}_2\right) = \prod_{i=1}^{k} \text{PCMR}_i \; .$$

The end-to-end time required to transmit the CROP from the central fusion facility to the users measures network currency. One approach to doing this is to calculate the average of all paths from source to user. The overall average network transmission delay is then the average of these times,

$$\bar{t} \, ,$$

so that the currency component of the transformation function is

$$Q_{cur}\left(\mathbf{F}_3 \mid \mathbf{F}_0, \mathbf{F}_1, \mathbf{F}_2\right) = e^{-\bar{t}} \; .$$

Shared Information

Shared information is an essential ingredient of effective collaboration, as Chapter Four will discuss more fully. We refer to the CROP received by the users as the *observed* CROP. Matrix $\mathbf{F}_2$ represents the *fused* CROP. Each user's observed CROP is a subset of the fused CROP. The overlap among these subsets constitutes the information shared among the users. Information not in the overlap has the potential to be shared through the process of collaboration. The ability to collaborate therefore has the potential to increase the amount of information shared among the users, thus contributing to shared situational awareness.

Chapter Four

THE COGNITIVE DOMAIN

The more we accurately search into the human mind, the stronger traces we everywhere find of the wisdom of Him who made it.

—Edmund Burke

In the information domain, the data collected on the physical domain are processed and disseminated to friendly users. The product produced is the observed CROP, consisting of a function of the set of feature vectors, F_0, F_1, F_2, and F_3. The quality of the information used to produce the picture depends on the functional architecture in each of the information subdomains. In the cognitive domain, the products of the information domain are used to take decisions. The mental processes that transform the CROP into a decision and a subsequent action are not well understood, beyond the fact that they follow a general progression from awareness (a person's cognitive, holistic view of the battlespace) to understanding (the extraction of meaning from the holistic battlespace view) to decisionmaking. These processes depend on a range of factors, a few of which are psychological.

Here we focus on awareness and shared awareness, the steps before understanding, decision, and subsequent action take place. With respect to the NCW Conceptual Framework, we are focusing on the awareness attribute subsets of the Quality of Individual Sense Making and Degree of Shared Sensemaking measures. The cognitive processes that transform the CROP into a decision and subsequent action must be described for participants in the decision process both as individuals and as interacting, collaborating members of a decisionmaking team. Continuing the Information Superiority Value Chain depicted in Figure 2.2 through to shared situational awareness, we get the transformation diagram depicted in Figure 4.1.

Our objective is to describe a modeling framework that begins to reflect the factors most likely to influence individual situational awareness, shared situational awareness, and collaboration prior to decisionmaking. This chapter emphasizes completeness measures and metrics. Our approach is to establish a theoretical foundation by developing definitions, metrics, and hypotheses regarding their relationships.

ANALYSIS IN THE COGNITIVE DOMAIN

The decisionmaker must possess some degree of *situational awareness* to *understand*, i.e., draw inferences from the picture presented to him. Regardless of how complete, correct, and timely the CROP presented to the commander, there is no guarantee that he will be *cognizant* of the relevant knowledge reflected in the CROP.[1] An important question, then, is how correct, complete, and current the information must be to ensure sufficient understanding for the various levels of awareness and, ultimately, an accurate assessment of the situation the commander is facing. The answer depends on a number of factors including the decisionmaker, the decision support system, and the collaborative environment. Consequently, it is

RAND *MR1467-4.1*

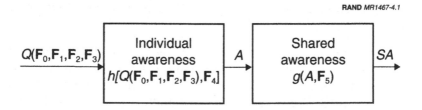

Figure 4.1—Cognitive Domain Transformations

---

[1]See Appendix A for definitions of the italicized terms.

important to develop a relationship between the various quality metrics and situational awareness as a function of the key parameters that characterize these ingredients.

Realizing the complexity of representing the functional relationships between the various quality metrics and awareness, we chose to address them in the following order:

- **Situational awareness of the individual.** First, model an individual decisionmaker's situational awareness, given key attributes characterizing individuals and information quality from the Information Domain;

- **The collaboration process for a team.** Next, model the role of information available to individuals participating together in some joint collaborative action; and

- **Shared situational awareness of the team.** Finally, modify the model of an individual's situational awareness to include factors representing information available to collaborating individuals, then extend this model to the situational awareness of an entire collaborating team.

MODELING INDIVIDUAL SITUATIONAL AWARENESS

Identifying what it will take—the level of information quality—for a decisionmaker to assess the situation presented to him correctly—to "be aware"—is complex. Several factors come into play: education and training, experience, the current situation, cultural background, personality, language, the opportunity to collaborate with others, the *quality of the information* presented, etc. We propose the following metric for individual situational awareness:

> **Degree of Situational Awareness:** the fraction of fused feature vectors the decisionmaker is capable of realizing.

Stated this way, this equates to the degree to which the decisionmaker is aware of the situation facing him. This metric emphasizes the use of the individual components of the CROP and includes a reference to the ability of the individual decisionmaker. It does not place greater or lesser value on the decisionmaker's correct realization of fused feature vectors; rather, it focuses exclusively on his ability to realize what he sees. It is essentially a completeness metric. Additional metrics (e.g., correctness of awareness) remain to be developed in future research.

Modeling the Individual

It is impossible to deal with *all* the factors that contribute to an individual decisionmaker's level of situational awareness in any coherent way. Instead, we chose to use an agent representation of a decisionmaker. That is, some combination of the factors listed in the previous paragraph will predispose the commander to grasp the situation quickly, and some combinations will not. In the absence of research in this area, we simply posited a continuum and selected discrete points on that continuum. For example, suppose we focus initially on education and training, experience, and familiarity with the current situation. For all these factors, the domain is clearly continuous. Rather than deal with the complexities of continuous domains, we instead defined two discrete points for each, as shown in Table 4.1.

A strict combinatorial assessment of the example attributes in Table 4.1 produces eight distinct possible awareness conditions that characterize a commander's predisposition to grasp the situation presented to him. We refer to these as *decision agents*. Suppose we focus on just four that descend in the order of awareness of the situation presented. These are described in Table 4.2, where the exemplar decision agents are denoted by Φ_i.

Next, we interpreted $A = h[Q(F_0,F_1,F_2,F_3),F_4] \in [0,1]$ to be the degree (level) of individual situational awareness each of the commander decision-agent types possesses. Recall that $Q(F_0,F_1,F_2,F_3)$ is the quality (completeness in this case) of the CROP produced in the information domain, and F_4 is the observed CROP. For example, a commander who is totally aware of the situation presented to him—one who is able to realize almost all the observed CROP—has a score close to 1.[2] The remaining question is how the quality of the information represented in the CROP influences the commander's awareness. It

---

[2]Note that total situational awareness does not equate to "willingness to decide." It may be that the commander is "aware" of the fact that he has insufficient information to act wisely—that is, insufficient information to make appropriate inferences about the situation.

seems reasonable to assume that, regardless of the commander's inherent awareness characterization, information of higher quality will tend to increase awareness over some range.[3] Consequently, we

Table 4.1

Exemplar Discrete Awareness Attributes

Attribute	High	Low
Education and training	Graduate of advanced service and civilian schools	Limited education beyond undergraduate studies and basic service school
Experience	Senior officer who has commanded troops in several operations	Junior officer with limited combat experience
Current situation	Familiar with the current situation	Unfamiliar with the current situation

NOTE: The entries in this table and in Table 4.2 are notional and are not the result of settled research. Clearly, several other attributes and combinations of attributes may be at play.

Table 4.2

Exemplar Decision Agents

Φ_i	Agent Characterization	Description
Φ_1	Highly capable	An experienced, well-educated commander who is familiar with the situation confronting him and a veteran of considerable field training
Φ_2	Less capable	An experienced commander with limited education who is unfamiliar with the situation confronting him but has some field training
Φ_3	Marginally capable	An inexperienced commander with limited education who is unfamiliar with the situation confronting him but has some field training
Φ_4	Incapable	An inexperienced commander with limited education who is unfamiliar with the situation confronting him and has little field training

---

[3]Clearly, this is another area in which additional research is needed.

sought a functional relationship in which the dependent variable is "degree of awareness," A, and the independent variables are information quality measures (completeness, in this case).

Individual Situational Awareness

The remaining chore is to relate the awareness of the alternative decision agents to each of the total information quality measures. This suggests isorelationships for each of the measures. Recall that the situational awareness range (the dependent variable) is $A \in [0,1]$. For completeness and correctness of situational awareness, the domain can also be between 0 and 1. What is needed is a relationship that shows decision agents with high awareness (Φ_1 and Φ_2) becoming markedly more aware of the situation with increasing completeness and correctness of information, and less so for decision agents with low awareness (Φ_3 and Φ_4). Figure 4.2 is a notional illustration of such a relationship for completeness.

The four curves are isofunctions whose shape depends on the ability of the individual decision agent as described by the parameter, Φ_i. Consequently we modify the transformation depicted in Figure 4.1 so that we have $A_{\Phi_i} = h[Q(F_0, F_1, F_2, F_3), F_4]$. The subscripted notation will be used from here on through the rest of the report. Figure 4.2 therefore depicts the degree of situational awareness that the four notional decision-agent capability levels attain as the quality (completeness) of the CROP varies.

The Effects of Fusion Levels

The level of situational awareness of an individual will likely be affected by the *level of fusion* in the information processing system supporting the decisionmaking process (e.g., the type of display and the format for presentation of the information). The levels of fusion were defined in Table 3.1. For example, the situational awareness curves for the same four decisionmakers depicted in Figure 4.2 might resemble those depicted in Figure 4.3 at a lower level of fusion. Suppose that Figure 4.2 represented awareness at fusion level 3, i.e., the fused CROP contains "assessments of enemy intent." Figure 4.3 might then represent awareness at fusion level 1, a simple set of "refined tracks and identity estimates."

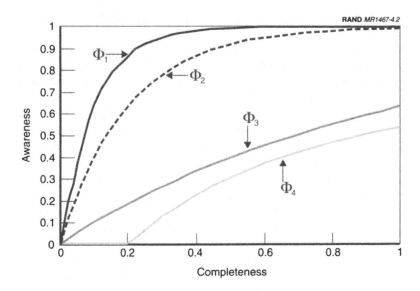

Figure 4.2—Notional Effect of Complete Information on Awareness

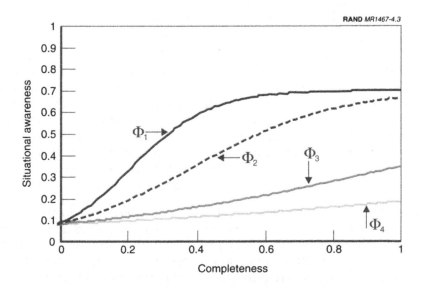

Figure 4.3—Fusion-Level Effects on Situational Awareness

MODELING SHARED SITUATIONAL AWARENESS

Individual awareness must now be refined to account for the effects of individuals participating in some joint action prior to decision-making. That is, we wish to explicitly represent the effects of *shared situational awareness* on the individual decisionmaker as the first step toward measuring the overall effects of shared situational awareness for a collective decision team. Our model of individual situational awareness thus far includes a representation of the differences among individual decision agents and the quality of information produced in the information domain. To describe shared situational awareness, we must augment the current model by representing the complex interactions in situations involving more than one individual. We propose the following metric:

> **Degree of Shared Situational Awareness**: the fraction of fused feature vectors in the CROP that members of a team realize similarly, whether or not they collaborate.

This metric emphasizes the importance of individual situational awareness and allows agreement to exist even when individual decisionmakers have not collaborated. Collaboration may not be needed when (1) information quality is very good, (2) each of the commander agents is highly capable and knows that the other commander agents are similarly capable, and (3) the situation at hand is not unusual or complex. Consider, for example, an experiment in which information is sent to many people. Each person individually interprets that information received. People are then interviewed to ascertain what they thought they knew about the situation. An outcome of this experiment could be that these persons have shared perceptions even though they did not collaborate. Their beliefs about what other persons thought or perceived may not have mattered.

The metric also considers the effects of collaboration during joint action. *Joint actions* are sequences of jointly performed acts occurring within some process over a period of time. A *joint act* is a single instance of a coordination action undertaken by two or more individuals. Joint actions require the coordination of what the participants intend to do and the physical and mental states required to execute those intentions (see Clark, 1996, p. 59). The metric we developed for degree of shared situational awareness emphasizes the importance of factors affecting individual situational awareness, the quality of collaboration, the influence of collaboration on individual decision agents (including the importance of belief about presuppositions, as this chapter discusses later), and factors representing situation complexity. The focus here will be on modeling the quality of collaboration and the influence of collaboration on individual decision agents.

We approached this modeling by formulating hypotheses about these complex interactions and their effects on shared situational awareness. We have not attempted to test these hypotheses in any rigorous way; rather, we assumed their validity and built models that are consistent with them. The exemplar measures and metric suggested in Appendix B follow from the models. Our research suggests that the hypotheses presented in the subsequent subsections collectively define a reasonable theoretical starting point for modeling shared situational awareness.

Collaborating Teams

We first hypothesized that, *when employed, collaboration is critical for determining shared situational awareness.* As mentioned earlier, under some circumstances, shared situational awareness could be achieved without collaboration. In addition, the effects of collaboration can be positive or negative. They are positive when they fill gaps in knowledge or confirm existing knowledge and are negative when gaps remain and evidence received disconfirms what is currently "known." In the latter case, time is required both to gather more information and to resolve inconsistencies. For now, however, we focus on assessing the important attributes that affect teams that *do* collaborate and therefore have either positive or negative impacts on the degree of shared awareness.

Two significant categories of attributes affect collaborating teams: *individual* and *group*. Table 4.3 lists some of the attributes in each category. Both individual and group attributes are important for describing collaboration; therefore, we considered both in modeling shared situational awareness. Of concern, however, is identifying the factors that affect interactions among collaborators and selecting those that have the most effect. The literature seems to suggest that

Table 4.3

Attributes Affecting Collaborating Teams

Characteristics	Attributes
Individual	Experience
	Familiarity with situations similar to the current one
	Ability to share knowledge
	Ability to access the knowledge of others
	Access rights
	Authority level
	Competence with the collaboration tools
Group	Task structure
	Role specification
	Shared operational model
	Degree of common language
	Group dynamics
	Quality of the interoperability the collaboration environment provides

formality of interaction, group size, group roles, and task complexity have the greatest effect.[4] We derived our structural model for collaborative team interaction from these attributes and factors.

Common Ground

Herbert Clark (1996, p. 12) has defined *common ground* as "the great mass of knowledge, beliefs, and suppositions [participants] believe they share." Robert Stalnaker (1978, p. 321) described *common ground* in terms of what he called "presuppositions":

> Roughly speaking, the presuppositions of a speaker are the propositions whose truth he takes for granted as part of the background of the conversation ... Presuppositions are what is taken by the speaker to be the *common ground* of the participants in the conversation, what is treated as their *common knowledge* or *mutual knowledge*. [emphasis added]

---

[4]Two others mentioned are: time and task dependency, and the physical location of collaborating team members.

Both definitions imply that some sort of joint activity is taking place and that *common ground* describes what the participants believe they share. In this context, "[A] joint action is one that is carried out by an ensemble of people acting in coordination with each other" (Clark, 1996, p. 3). Joint activities can be represented by the sequence of states attained by the joint activity and the trace of moves or actions that caused the state changes. During a joint activity, common ground accumulates among participants.

We adopted a version of Clark's definition (see Appendix A). This suggests that, at any moment in a joint collaborative process, common ground consists of the three components listed in Table 4.4, along with the corresponding information attributes.

Collaboration and Common Ground

A second hypothesis emphasizes the importance of common ground to the collaboration process. *To be effective, collaboration requires both the development of common ground among collaborators and familiarity with the capabilities of other collaborators.* We consider two varieties of common ground: *actual* and *assumed.* Because

Table 4.4

Common Ground Components and Attributes

Component	Definition	Information Attributes
Initial common ground	The background facts, assumptions, and beliefs participants presuppose at the start of a joint activity	Information that is officially part of the joint activity
Current state of the joint activity	The presuppositions each participant has about the current state of the activity	Information that is not officially part of the joint activity
Public events so far	The events participants presuppose have occurred in public leading up to the current state	

common ground is defined by what people *think* they share, there may be differences between what they *think* they shared and what they *actually* share. The transition process from assumed common ground to actual common ground may not be uniform for all group knowledge, beliefs, and suppositions.

Common ground does not develop instantaneously when there is collaboration. Clark approached the development of common ground through the passage of states in joint actions. Morris Friedell (1967, pp. 28–39) introduced a Boolean algebra model of common ground. He noted that there is a period of "Initial Calibration" during which participants "tune in" to each other and move from a state of common sense to states of common opinion and common knowledge. Friedell also based his analysis on what each participant believes about himself and about others, defining *belief* as "a proposition that a participant would assent to if given ample opportunity to reflect" (Friedell, 1967, p. 2). *Common opinion* represents perceived symmetry of belief between two or more individuals. *Common knowledge* is true common opinion and is achieved when "common knowledge implication" exists. In this state, who knows what about subject x and who knows what about subject y are considered to be common knowledge. This is a critical change from common ground, in that it assigns specific responsibility for certain knowledge to specific individuals and states that these responsibilities are known among all collaborating participants.

These concepts—"initial collaboration" and "common knowledge implication"—lead us to our next step in modeling shared situational awareness: defining a model for the storage and retrieval of opinion and knowledge within a collaborating team. This model is also the vehicle for describing the transition from assumed to actual common ground.

Transactive Memory Systems

Freidell's "common knowledge implication" concept suggests that it is important to know who knows what about a particular problem. A positive outcome of collaboration, if it occurs, may be that shared facts previously known only to distinct participants can provide for interpretations that would not have been possible otherwise.

The states of a joint action represent the succession of results of joint acts. Traces represent the process steps executed between two successive states. Clark notes that the current state of an activity is often carried in an external representation, such as a chessboard. He also notes that discrepancies develop between the individual representations of common ground that each participant in a joint action maintains. These discrepancies are especially important to the discussion of coordination problems—where two or more people have "common interests, or goals, and each person's actions depend on the actions of the others" (Clark, 1996, p. 62). The coordination devices for such problems are clues for coordinating behavior or a key that is mutually recognized as such. Representation of these clues and keys is the next step. One structural model for defining and analyzing "common knowledge implication" is Daniel Wegner's transactive memory system, which he defined to be "a set of individual memory systems in combination with the communication that takes place between individuals" (Wegner, 1987, p. 186). He also states that the "study of transactive memory is concerned with the prediction of group (and individual) behavior through an understanding of the manner in which groups process and structure information" (Wegner, 1987, p. 185). The transactive memory system model addresses the storage and movement of information among actors, essentially recording the artifacts of the states and traces that define the processes of joint actions in Clark's model of common ground. This suggests the following corollary to the common ground hypothesis:

> The performance of a collaborative team is strongly dependent on the completeness of its system for recording and retrieving information.

The transactive memory system appears to be a reasonable model of the primary factors affecting collaboration. It is consistent both with Clark's model for common ground and with cognitive structures in social cognition theory (Pryor and Ostrom, 1987).

Below, we will first address the structure and contents of a transactive memory system for shared situational awareness. Then we propose a process model that captures the development of the structure and contents of a transactive memory system for shared situational awareness.

Structure and Contents of the Transactive Memory System

An individual can store information internally and retrieve it according to his own encoding, storage, and retrieval processes. An individual's *metamemory* includes knowledge about these processes. If an individual stores information externally, the storage and retrieval process must also include the location of the information. A person must remember whether a particular memory item (label) is stored internally or externally (location) to be able to access the memory item's value (the item).

If externally stored information resides in another person, a transactive memory system exists. Individuals can be assigned as information stores because of their personal expertise or through circumstantial knowledge responsibility. Labeled items may be uniquely assigned to participants in the transactive memory system (a differentiated system), or many individuals may store subsets of the same labeled items (an integrated system).

Figure 4.4 depicts the components of a transactive memory system for shared situational awareness. Each individual participating in the transactive memory, I_j, has a set of memory components (a, e). These memory components capture the key elements of the collaboration. Clark's concepts of communal common ground—embodying shared expertise and including communal lexicons, cultural facts, norms, and procedures—is consistent with this transactive memory model. His concepts of personal common ground—including perceptual bases, personal diaries, and personal lexicons—are also consistent with this transactive memory model.

The shared information CROP memory component represents the portion of the CROP received by individual I_j. It is possible that not all individuals receive the complete CROP; therefore, only a subset of the observed CROP may be common to all individuals. The private-role or situation-specific information memory component represents the individual's internally stored memory encodings (labels) and memory items (information), not shared with other participants during collaboration, that affect the individual's ability to interpret elements of the CROP. These memory items may not be shared because of time or other collaboration environment constraints.

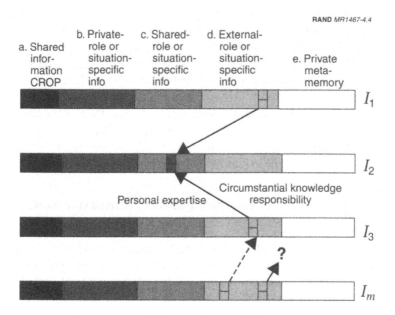

**Figure 4.4—Transactive Memory System for Shared
Situational Awareness**

The shared-role or situation-specific information and the external-role or situation-specific information components are the core of the transactive memory system. These components represent information that some individuals store externally in other individuals and that some individuals retain on behalf of other individuals in the transactive memory system. There can be direct links between an individual and the retrieval of a memory item (e.g., the link between I_1 and I_2). There can be indirect links that take "hops" through the transactive memory system until the memory item is accessed (e.g., the link between I_m and I_2). There can also be retrieval failures in a transactive memory system, for example, when an individual to whom external storage was assigned departs the transactive memory system or when an individual is temporarily incapacitated so that he forgets where information is stored. The private metamemory component contains each individual's knowledge about the processes of encoding, storage, and retrieval for his or her individual memory and for the external memory in the transactive memory system.

The effects of automation can be assessed using Figure 4.4. Component *a* of an individual's transactive memory is affected by automation, the effects of which have been addressed in the production of the CROP. Automation does not affect components *b* and *e* of an individual's transactive memory. Automation can, however, affect components *c* and *d* of an individual's transactive memory. Components *c* and *d* may be developed through at least three mechanisms: nonautomation means, automation support sharing, or previous transactive memory systems developed by other individuals and shared with the current individual.

Process Model for Development of the Transactive Memory System

Ulhoi and Gattiker (1999) defined an iterative process for incremental knowledge divergence and convergence in describing how people develop a conceptual framework for a technological problem. Figure 4.5 applies this iterative refinement to development of the transactive memory system for shared situational awareness.

The key features of this process are the iterative stages of individual information assessment, followed by team discussion, leading to some state of shared situational awareness. The team discussion period consists of reinforcing and refuting current beliefs about the situation. Knowledge divergence results from the presentation of interpretations from collaboration team members. Knowledge convergence is the result of consensus derived from assessments about the beliefs of other team members during the team's discussions. As the extent of prior collaboration increases, team members not only pass through the previously identified iterative stages, they also transition from assumed to actual (confirmed) beliefs about what is known and shared by others in the team. Assumed beliefs dominate the initial calibration state. Confirmed beliefs dominate the common knowledge state.

Figure 4.5 depicts the three key shared situational awareness states in this process. In the initial calibration state, team members have achieved consensus about what roles each team member can and will perform in the collaboration, and the team generates initial alternative actions for further discussion. In the structured knowl-

RAND *MR1467-4.5*

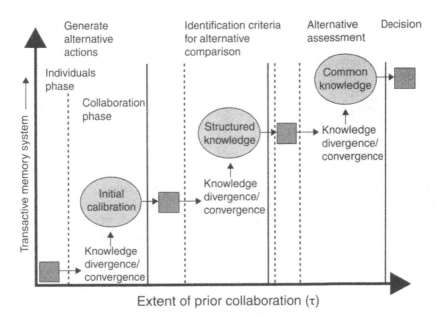

**Figure 4.5—Process Model for Developing the
Transactive Memory System**

edge state, team members have begun to organize internally and
externally stored information for the situation, and the transactive
memory is forming. The team has progressed to consensus about
who knows what about the situation and identifies alternatives for
further analysis and comparison. In the final state, common knowledge, not only do team members reach consensus about who
believes what but also about what is true for the situation.
Alternatives proposed are assessed against the consensus reached
about who knows what and about what is true.

No specific models or metrics are associated with this process. The
process is used as a reference framework for interpretation of results
obtained from other models proposed for shared situational awareness.

Modeling Familiarity

In his dissertation on KC-135 crew performance at Fairchild Air Force Base, Washington, Daryl Smith (1999) used the term "crew hardness" to measure how often crew members fly as an integral team. We adopted the spirit of Smith's measure and use the term "team hardness" in our model of shared situational awareness. Figure 4.6 depicts the rotation of ten individuals in and out of teams during some period, τ.

As participants develop stronger relationships with other participants through repeated or continued team interaction, the links between the participants become stronger. Clark refers to this, in the language of common ground, as the establishment of team *grounding*. From this, we hypothesized that the strength of these links is related to the degree of common ground between participants. This suggests a second corollary to the common ground hypothesis: *The completeness of the system for recording and retrieving information depends on how frequently the team has recently collaborated.*

The links, though intuitively appealing, summarize many of the attributes and interaction effects set out at the start of this discussion. We required a more definitive model for strengthening links between individuals as team hardness increases, as the links in Figure 4.6 depict. The structure of the transactive memory system then allowed us to define a measure for team hardness.

RAND *MR1467-4.6*

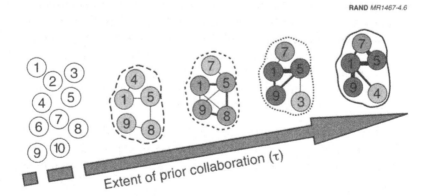

Figure 4.6—Collaborative Team Development

Estimating Team Hardness

The preceding discussion provided an intuitive definition of team hardness: As participants develop stronger relationships with other participants through repeated or continued team interaction, the links between the participants become stronger. The transactive memory system contains explicit links capturing the amount of shared information between team members. Figure 4.4 depicted example links between individuals I_1 and I_2 and between individuals I_m, I_3, and I_2. A metric that measures the complexity of those links, the number of "hops" required to retrieve information, is desirable. We developed estimates of link complexity by referring to the labels assigned to the transactive memory system components for individual Is in Figure 4.4. If the number of externally stored items in the transactive memory system were measured and tracked as the team hardens, we hypothesize that

1. *The ratio of the number of shared-role or situation-specific (c) and external-role or situation-specific information (d) to private-role or situation-specific information (b) elements would increase as the transaction memory system becomes complete, where completeness is relative to the complexity of the situation.* The resulting mathematical expression is

$$TM_{complex} \approx \frac{c+d}{b} \, .$$

2. *The ratio of "hops" to the number of shared-role or situation-specific (c) information elements in the transactive memory system would indicate the complexity of the transactive memory system.* The resulting mathematical expression is

$$TM_{complex} \approx \frac{hops}{c} \, .$$

It is possible to collect information indicating these ratios through analysis of transactions and communications among individuals participating in team exercises, as the team exercise progresses. This task is left to further research.

These two hypotheses served as the basis for our functional models of team hardness. These models are time dependent. If t represents the time elapsed since the start of the operation and τ represents the length of time the team has been training or operating together, $TM(T)$ represents transactive memory, where $T = \tau + t$.[5] As with previous completeness metrics, the models selected for transactive memory have the property $0 \le TM(T) \le 1$.

With a model for the transactive memory system and a description of its completeness, it is next necessary to define a process in which the transactive memory system is developed by a collaborating team. To do this, it is first necessary to understand the role of *consensus* among the team members.

Consensus

Consensus plays a central role in developing a transactive memory system. It is the majority *opinion* of a team arrived at through active collaboration. Its definition *implies the existence* of shared situational awareness. Consensus was the engine driving the knowledge divergence-and-convergence cycles presented in Figure 4.5. Knowledge convergence is achieved by consensus about team members' beliefs during the team's discussions. Understanding the nature of beliefs and how they can be represented in a form that allows reasoning about the interactions of persons participating in decisionmaking teams is therefore essential to our understanding of collaborative decisionmaking and the role of consensus. Consensus is an example of a degree of shared situational awareness. Noting that not all collaborating individuals have to agree before a decision and subsequent action can take place, we are interested in a measure of the degree of consensus. We hypothesized that *the degree of consensus can be estimated by the number of pairwise combinations of collaborating individuals who interpret feature vectors similarly*.[6]

It is possible for consensus to exist on a proposition that is "correct" as well as on one that is "incorrect." To complete the foundation for the modeling of shared situational awareness, it is necessary to frame

---

[5]Appendix B includes an example of this.

[6]Appendix B includes an example mathematical statement of this measure.

consensus about truths and about untruths. Friedell's approach to this issue is to differentiate opinion and knowledge (Friedell, 1967). Common *opinion* represents the common belief of a team; common *knowledge* speaks to the truth or untruth of that belief.

Common opinion and common knowledge can exist with or without collaboration. If all commander agents expect all other commander agents to interpret a fused feature vector a particular way, common opinion and common knowledge may develop without explicit collaboration. Our model captures the distinction between "*A* knows *x*" and "*A* believes *x*, and *x* is true." Taken together, these observations support the hypothesis that *collaboration does not imply consensus*.

We used the compositions of decisionmakers' beliefs, as represented by their interpretation of fused feature vectors and defined by their degree of individual situational awareness, A_{Φ_i}, to model common opinion in shared situational awareness.

Shared Situational Awareness

Our models of shared situational awareness integrate the modeling we proposed earlier. First, we placed the individual in a team and measured his situational awareness in a team setting. Note that this is not the same as team awareness but rather the effect of team dynamics on an individual member of a collaborative decisionmaking process. The contribution is essentially derivative of the transactional memory function and, therefore, of team hardness. Second, we addressed the consensus that develops among collaborating individuals and its affect on the team's shared situational awareness. Finally, we accounted for diversity of decision agent capabilities among the collaborators to create a composite model for the degree of shared situational awareness.

Team participation can have both salutary and deleterious effects on individual awareness. Presumably, team participation produces positive synergies that improve individual performance. However, in some instances, individual team members with limited ability but with positions of authority might enforce their will on the process, to the detriment of other individuals in the team. In addition, the fusion subdomain metrics indicate that quality, $h[Q(F_0,F_1,F_2,F_3),F_4]$, may decrease over time. This effect is included in the individual shared awareness function, developed earlier.[7] A revised measure of individual situational awareness that combines these factors and that includes transactive memory, $A_{\Phi_i}(t)$, needed to be developed. The revised, time-dependent $A'_{\Phi_i}(t)$ is the fraction of the observed feature vectors the individual decisionmaker with capability Φ_i realizes with the benefit of team participation, when the operation has gone on for t time units. It is therefore taken to be the completeness of the CROP realized by agents of capability Φ_i at time t.[8]

The next step was to evaluate the situational awareness of the team when working together, $SA(t)$, recognizing that it too will be time dependent. This is what we refer to as *shared situational awareness* and is depicted without the time dependence in Figure 4.1. In general,

$$SA(t) = g\left(A'_{\Phi_i}, F_5\right),$$

the collective fraction of the observed CROP, $Q(F_0, F_1, F_2, F_3)$, that can be interpreted by the entire team, is our representation of the degree of consensus. In this formulation, F_5 represents the observed CROP realized by agents of capability Φ_i at time t. This will be a function of the individual situational awareness of the members with different capabilities when working in the team environment and their situational awareness of the individual feature vectors in the observed CROP. For any two team members, for example, we wish to know which feature vectors they can jointly interpret. This implies that we know not only the fraction of the observed CROP that they can interpret but also *which* features they can interpret. Appendix B develops an example metric based on the pairwise assessment of the interactions of team members.

SUMMING UP

In the information domain, the data collected on the physical domain are processed and disseminated to friendly users. In the cognitive domain, the products of the information domain are used

---

[7] Appendix B includes an example.

[8] Appendix B develops an example function with the appropriate characteristics.

to take decisions. The mental processes that transform the CROP into a decision and a subsequent action depend on a range of factors, a few of which are psychological. This chapter focused on the steps before a decision is made and subsequent action takes place. The cognitive processes that transform the CROP into a decision and subsequent action must be described for participants in the decision process, both as individuals and as interacting, collaborating members of a decisionmaking team. We described a modeling framework that attempts to reflect the factors most likely to influence individual situational awareness, shared situational awareness, and collaboration prior to decisionmaking.

Modeling Individual Decisionmaking

Several factors influence what it will take for an individual decision-maker to assess the situation presented to him correctly. Among these is the quality of the information presented. The metric we developed, the degree to which the decisionmaker is aware of the situation facing him, emphasizes the use of the individual components of the CROP and includes a reference to the ability of the individual decisionmaker. It is interpreted to be the fraction of the observed CROP the decisionmaker realizes.

We suggested an agent representation of a decisionmaker based on combinations of decisionmaker capability attributes. We defined two discrete points for each attribute and generated four decision agents possessing these attributes at one of the two points. Consequently, we suggested a functional relationship in which the dependent variable is "degree of awareness" and the independent variables are information quality measures (completeness in this case).

The end result of this process is an explicit relationship between the quality of the observed CROP and the ability of the decisionmaker:
$A_{\Phi_i} = h[Q(\mathbf{F}_0,\mathbf{F}_1,\mathbf{F}_2,\mathbf{F}_3),\mathbf{F}_4]$.

Modeling Shared Situational Awareness

Individual awareness was refined to account for the effects of individuals participating in some joint action prior to decisionmaking, thus explicitly representing the effects of *shared situational awareness* on the individual decisionmaker. To describe *shared situational awareness*, we augmented the current model by representing the complex interactions in situations involving more than one individual. The metric we proposed is the fraction of fused feature vectors in the observed CROP that members of a team realize similarly, whether or not they collaborate. This metric emphasizes the importance of individual situational awareness and allows agreement to exist even though individual decisionmakers have not collaborated.

We hypothesized, however, that when employed, collaboration is a critical factor in determining shared situational awareness. We focused on assessing the important attributes that affect teams that *do* collaborate and that therefore have either positive or negative effects on the degree of shared awareness.

The concept of common ground was introduced as an ingredient in the shared situational awareness process. We took it to mean the knowledge, beliefs, and suppositions team members believe they share. During a team activity, therefore, common ground accumulates among team members.

We further hypothesized that to be effective, collaboration requires both the development of common ground among collaborators and familiarity with the capabilities of other collaborators. Common ground does not develop instantaneously when there is collaboration, there is a period of "initial calibration" during which participants "tune in" to each other and move from a state of common sense to states of common opinion and common knowledge.

A structural model for defining and analyzing this phenomenon is a *transactive memory system*, defined as a set of individual memory systems in combination with the communication that takes place between individuals. This model is concerned with the prediction of group (and individual) behavior through an understanding of how groups process and structure information.

An individual can store information internally and retrieve according to his or her own encoding, storage, and retrieval processes. If an individual stores information externally, the storage and retrieval process must also include the location of the information. If externally stored information resides in another person, a transactive memory system exists. Individuals can be assigned as information stores because of their personal expertise or through circumstantial knowledge responsibility. Each individual participating in the transactive memory has a set of memory components. These memory components capture the key elements of the collaboration. They represent information that some individuals store externally in other individuals and that some individuals retain on behalf of other individuals in the transactive memory system. The links between an individual and retrieval of a memory item may be either direct or indirect, taking "hops" through the transactive memory system until the memory item is accessed.

As participants develop stronger relationships with other participants through repeated or continued team interaction, the links between the participants become stronger. This suggested a second common ground hypothesis: The completeness of the system for recording and retrieving information depends on how frequently the team has recently collaborated. This concept was referred to as "team hardness."

A time-dependent functional model for team hardness was defined as $0 \leq TM(T) \leq 1$, and $T = \tau + 1$, where t represents the time elapsed since the start of the operation, and τ represents the length of time the team has been training or operating together. An example is included in Appendix B.

Consensus plays a central role in developing a transactive memory system. It is the majority opinion of a team arrived at through active collaboration. Its definition implies the existence of shared situational awareness. Noting that not all collaborating individuals have to agree before a decision and subsequent action can take place, we are interested in a measure of the degree of consensus. We hypothesized that the degree of consensus can be estimated by the number of pairwise combinations of collaborating individuals who interpret feature vectors similarly.

Models of shared situational awareness integrate our earlier modeling. First, we placed the individual in a team and measured his situational awareness in a team setting. Note that this is not the same as team awareness but rather the effect of team dynamics on an individual member of a collaborative decisionmaking process. The contribution is essentially derivative of the transactional memory function and, therefore, of team hardness. Second, we addressed the consensus that develops among collaborating individuals and its influence on the team's shared situational awareness. Finally, we accounted for the diversity of decision agent capabilities among the collaborators that results in our composite model for the degree of shared situational awareness.

Implications

The foregoing analysis indicates that for a given team, the time to accomplish a task using collaboration decreases as team hardness increases. This squares with intuition and is further supported by Fred Brooks' assertion that the time to complete a software project varies with the degree of interaction necessary during the project (Brooks, 1995, pp. 15–19). The more interaction required, the longer the process, and therefore the greater the need for an experienced team, as Figure 4.7 illustrates.

For a fixed team size of n, the figure illustrates what happens when teams that possess varying levels of experience (hardness) tackle tasks with complex interrelationships. As the number of people in

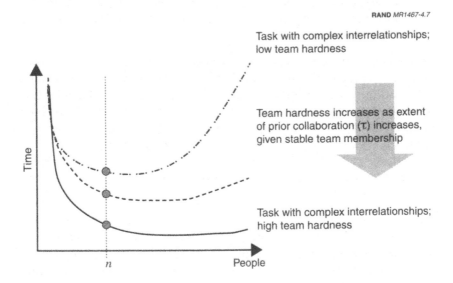

Figure 4.7—Team Size and Hardness Determine Task Duration

the team increases, the time required to complete the task decreases, but at varying rates. At some point, the time required for the teams to complete the task increases when additional people are added to the team. As with the decreasing rate, the increasing rate varies with the team's experience.

Teams with assumed common ground or assumed common knowledge require additional interaction to transition to actual common ground or actual common knowledge, as Figure 4.5 illustrated. Teams that are not hardened, as Figure 4.6 illustrated, have incomplete transactive memory systems and simple, rather than complex, transactive memory systems. Given tasks with complex interrelationships, a team with a low degree of shared situational awareness will require longer to complete the tasks than a team with a high degree of shared situational awareness. Given that this chapter focused on the cognitive process that transforms the CROP into a decision and subsequent action, we note that teams with low degree of shared situational awareness may be unable to reach the decision point in the time available. Clark and Brennan (1991) assessed the costs of grounding, noting 11 factors that contribute to achieving common ground. Further research to relate these proposed costs of grounding to the time taken for a team to accomplish a task, given its degree of shared situational awareness, is warranted. Also, the potential influence of training on how to collaborate and enabling information technology needs to be examined.

Chapter Five

FUTURE WORK

It is error only, and not truth, that shrinks from inquiry.

—*Thomas Paine*

The preceding chapters described a mathematical framework that might be used to develop detailed mathematical quantities that represent what are generally considered to be qualitative concepts. In some cases, data may exist in the military C^4ISR community to confirm or disconfirm both the process and the examples presented in Appendix B. In these cases, it will be necessary to locate and assess the data. Where data do not exist, further experimentation or historical analysis will be necessary.

As noted earlier, the discussion of the cognitive domain is not complete. The relationship between information quality and situational awareness is the first step in the decisionmaking process. Further work is needed to codify the relationship between situational awareness and the ability of the decisionmaker to make inferences from the CROP, that is, his understanding of the situation.

REFINEMENT OF CURRENT RESEARCH

The results reported in this document are not complete. The examples used to illustrate the process in Appendix B are notional and, although intuitively consistent, unsupported. The mathematical framework also needs testing as more evidence-supported metrics are developed. Finally, we have focused almost exclusively on the completeness measure of information quality. The same emphasis needs to be placed on correctness, currency, and an aggregate measure. In addition, more work is required to reflect the influence of collaboration on shared situational awareness adequately.

Data Fitting

The objective of this task is to use existing data to either confirm the validity of the relationships suggested in this work or to suggest different relationships. The C⁴ISR community could be polled for data that was generated experimentally, through exercises, or from pertinent military operations. Standard curve-fitting techniques and/or statistical "goodness-of-fit" analyses could be used, where applicable, to develop new metrics within this report's mathematical framework. Of particular interest would be collecting cognitive domain data that could be used to gauge the functions described in the situational awareness portions of the framework.

Experimentation

Another approach is to conduct experiments to gain additional insights about information quality and awareness. Experimentation is attractive because several parameters can be controlled. For example, decisionmakers can be selected who have various combinations of the awareness characteristics described in Table 4.1, or some set of characteristics that reflect recent research more closely. Familiarity with the situation can be controlled, as can collaboration. Experience is a function of the combat command positions held. The problems might be in the form of situation assessments (CROPs) presented to the participants in various ways. The degree to which participants are able to realize enemy intent from what is presented would then indicate their level of awareness.

Decision, Understanding, and Action

The links between awareness and understanding and between understanding and decision must be established. The level of awareness affects the ability to understand, i.e., to draw inferences about the CROP, such as enemy intent. The inferences in turn affect the decision to be taken and therefore the subsequent actions ordered.

The work presented in this report could be taken a step further. As stated at the beginning, the overall goal is to establish a relationship between information and combat outcomes. We postulated that high-quality information, relevant to the current situation, is the key to good decisions. The C⁴ISR system, from sensors to fusion processing facilities and algorithms, produces the needed information. Making the connection between information and combat outcomes therefore provides an excellent tool for measuring the added value of improved C⁴ISR systems. Doing so will require modeling the links not just from information to awareness (as done in this report), but from awareness to understanding and from understanding to decision.

Historical Analyses

Past battles are an important source of information on the value of information. A considerable amount of data is available from various sources that can provide insights into the relationship between the quality of information and the level of awareness. For several past battles, information concerning "what each side knew" is available in the form of intelligence reports. The quality of the information represented in the reports—as measured by timeliness, completeness, and correctness—could be assessed, given that ground truth is known. Since the outcome of the engagements is also known, the relationship between the quality of information and combat outcome can be evaluated.

Gaming

It is possible to use game theory to illustrate certain effects based on information imbalances between two opponents. Recent RAND research illustrated the effects of information superiority using a simple two-sided zero-sum game in which varying numbers of strategies were available to both sides (Darilek et al., 2001). A series of four games were run, with combat power the same for both sides. The relative information available to each side was varied in each case. The outcomes illustrated the considerable advantage gained by the side with information superiority.

The same type of analysis could illustrate the effects of quality of information on the level of awareness. In a two-sided game, each side strives to obtain high-quality information, in the sense that it should be complete, correct, and timely. At the same time, each side attempts to ensure that the opponent's information is of low quality. This can be achieved through various deceptive practices and through direct action against information-gathering and -processing assets. Several pairings of players with varying awareness character-istics are then played against the various information quality levels. The minimax criterion could be used to evaluate the outcome of the game, thus establishing a link from information quality to awareness to decision and, finally, to outcome.

APPLICATION OF THE RESEARCH TO OTHER C⁴ISR ARCHITECTURES

Another area for future research would be to apply the mathematical framework in this report to C⁴ISR architectures beyond the one we have discussed. The architecture we have described here is a largely automated system in which sensors automatically forward data to a centralized fusion array, which then pushes observed CROPs to the users. Analyzing other proposed architectures, such as posting and retrieving systems (in which sensors and users decide what data to post and what to receive), and broadcast and subscription systems (in which all data are broadcast over the network, and different nodes and users decide what types of data to "tune in") would be of significant interest.

SOME DEFINITIONS

> All arts acknowledge that then only we know certainly, when we can define; for definition is that which refines the pure essence of things from the circumstance.
>
> —*John Milton*

This appendix records the formal definitions of important terms and concepts used in the report. Not all the definitions listed are agreed to by the analytic community involved in the study of C⁴ISR processes and procedures. In particular, such terms as *information superiority*, *data*, and *knowledge* lack agreed definitions. To the extent possible, however, we have adopted standard definitions. However, the task of the ISMWG is to develop accepted definitions for all the terms used in this report, and to the extent that their definitions have been settled, we have adopted them here.[1] Other definitions serve the purpose of ensuring internal consistency.

MEASURES AND METRICS

This research focused on measures and metrics of information quality in the decisionmaking process; it is therefore important that we include definitions of both. At one level, measure theory offers strict mathematical definitions for measures and metrics that rely heavily

_____

[1] The NATO Panel on Best C⁴ISR Practices has proposed definitions for some of these terms. In addition, Joint publications and the Institute of Electrical and Electronics Engineers have offered definitions in various publications referenced in the text.

on an understanding of mathematical analysis (see, for example, Halmos, 1950). Norman Campbell suggested a less formal but more philosophical approach. In a 1921 book called *What Is Science*, he argued that the "measurable properties of a body are those which are changed by the combination of similar bodies."[2] For example, the weight and cost of a bag of potatoes will change if combined with a second bag. However, the variety and cooking qualities of the potatoes will not. For Campbell, then, "measurement is the representation of properties by 'numbers.'" In this work, however, we offer operational definitions that are more useful to our purpose. In practice, measures and metrics are sometime used interchangeably, and, in many instances, this causes no problems. However, we chose to distinguish between the two and therefore illustrate the difference between them and also how they are related.

For this work, we adopted a practical and simple definition for *measure*: a basis or standard of comparison. By this definition, for example, cost, distance, power, and brilliance are measures in that they are standards by which alternatives might be compared. Measures are not always characterized by a single standard; they may consist of more-complicated expressions for which the associated yardstick is not clear, such as "the degree to which all unit/target features are distributed to all users." The yardstick (metric) used to compare alternatives based on this measure must consist of some assessment of "degree" (percent, for example) and a clear definition of what is meant by "distributed to all users."

A *metric* is a scale used to assess an alternative's position for a given measure. A metric is generally defined on the real number line— dollars, feet, watts, candlepower, etc. The mathematical concept is that of a "distance" or the amount of separation between two "points" or alternatives. For example, if x_1 and x_2 are points on the real number line, then the distance between the two is $|x_1 - x_2|$. If x_1 and x_2 are nonnegative and if $x_1 > x_2$, the distance also measures how much further x_1 is than x_2.[3] By convention, metrics in this report vary from 0 to 1, with 1 equal to the "best" value and 0 to the "worst."

---

[2]The selection quoted above is taken from excerpts reprinted in Neuman (1988), p. 1,772.

[3]A formal axiomatic definition of metrics and metric spaces can be found in Kreyzig (1978).

INFORMATION

The following terms are fundamental to a theory about how information affects human and organizational behavior. Although the following are attempts to record precise definitions, we recognize that many of these terms have no agreed-on operational definition as yet. However, to the extent possible, we selected definitions that are consistent with the information "primitives" discussed in Alberts et al. (2001).[4] We defined these terms as follows:

- *Data*: any representation, such as characters or analog quantities, to which meaning might be assigned. Examples include radar returns, infrared detections, and sensor reports.

- *Information*: data that have been processed in some way to put the individual data elements into some meaningful context. One example would be the transformation of a radar return from a GMTI sensor into the location and velocity of a moving vehicle.

- *Knowledge*: conclusions that are drawn from the accumulated information based on patterns or some other basis for gaining insights constitute *knowledge*. For example, we "know" a large armored unit is traveling toward our position in a westerly direction because of GMTI returns and intelligence reports.

- *Situational awareness*: a realization of the current situation constitutes *situational awareness* and results from the interactions between prior knowledge, beliefs, and the current perception of reality. For example, it is possible to infer that the enemy will attack from the west based on (1) current knowledge that a large armored unit is traveling toward us in a westerly direction; (2) prior knowledge of the enemy commander's behavior under similar circumstances; and (3) belief, arising from years of training, that similarly configured, rapidly moving armored columns lead to an enemy attack.[5] In the text, *awareness* and *situational awareness* are sometimes used interchangeably.

---

[4]The complete set of primitives consists of sensing, information, knowledge, awareness, understanding, decisions, actions, information sharing, shared knowledge, shared awareness, collaboration, and synchronization. See Alberts et al. (2001).

[5]Other definitions of *situational awareness* also focus on a state-of-mind definition. Carl Builder referred to situational awareness as a state attained by a decisionmaker in

- *Understanding*: the ability of humans to draw inferences about the possible current and future consequences of a situation—the process of comprehending, of appreciating the meaning of a word, sentence, event, proposition, map display, etc. Current and past knowledge lead to inferences about the current situation, while understanding how events might unfold is a function of an awareness of the situation. When attempting to assess the situation, understanding produces a higher level of inferential capability and is sometimes referred to as sense making.

Clearly, the boundaries between each of the terms defined above are situation dependent. For example, according to the definition above, a track established through several observations of a moving column is information. However, if it is one of several tracks constituting a total enemy force, it may be a single data point in the friendly commander's attempt to estimate when and where the enemy force will attack.

INFORMATION SUPERIORITY

Joint Publication 1-02 formally defines information superiority to be: "[t]hat degree of dominance in the information domain which permits the conduct of operations without effective opposition" (DoD, 2003, p. 255). However, information superiority is really more than that because it closely aligns with information requirements. Asymmetric information needs are as much a part of determining information superiority as the ability to collect, process, and disseminate an uninterrupted flow of information is. Moreover, information is never constant on a high-tempo battlefield, and needs will therefore vary.

The official definition therefore requires answers for two important questions:

---

which he is cognizant of the key physical, geographical, and meteorological features of the battlespace that will enable his command concept to be realized (Builder, Bankes, and Nordin, 1999, p. xv). Some researchers have referred to this broader definition as *sense making*, thus emphasizing the close link between situational awareness and "understanding." See the glossary for a definition of the latter, as set forth by Smith (1999).

- What are the relative information needs of both sides?
- Under what conditions do those needs change?

From a practical standpoint, what is needed is *knowledge* of what each side tries to collect, process, and disseminate and how this will change as the battle progresses. The definition therefore requires knowledge of the collection, processing, and dissemination regimes for both sides. If the friendly commander is able to (1) collect, process, and disseminate information when needed; (2) understand how his adversary's information needs will change over time; and (3) stop the enemy from collecting, processing, and disseminating information, he can claim information superiority. This means that both sides can be expected to engage in information operations. Both sides must be capable of physically and electronically protecting their C⁴ISR assets and can be expected to concentrate some of their resources on attacking the other's C⁴ISR assets. This all suggests an expanded definition of information superiority:

> *Information superiority:* A side has information superiority when it is able to (1) collect, process, and disseminate information as needed; (2) anticipate the changes in its own and its adversary's information needs; and (3) deny its adversary the ability to do the same.

NETWORK-CENTRIC WARFARE

It is impossible to ignore the compelling arguments made for the contribution of NCW to combat outcomes. The dual nature of the concept, networks and improved command and control procedures, makes it central to any discussion about the quality of information and decisionmaking—particularly when assessing their influence on combat operations.

NCW can be defined, according to Alberts, Garstka, and Stein (2000, p. 2), as

> an information superiority–enabled concept of operations that generates increased combat power by networking sensors, decision makers, and shooters to achieve shared awareness, increased speed of command, higher tempo of operations, greater lethality, increased survivability, and a degree of self-synchronization. In essence, NCW translates superiority information superiority into

combat power by effectively linking knowledgeable entities in the battlespace.

In contrast to network-centric operations or warfare, traditional warfare is considered to be *platform centric*. The difference between the two is that, in platform-centric warfare, one must mass force to mass combat effectiveness by physically massing platforms (weapon systems) to achieve combat effects; in NCW, combat effects are achieved by virtually massing forces through a robust network. That is, the employment of weapon systems on the battlefield is optimized so that a target is serviced by the most effective system in the network. Thus, the hypothesis is that a much smaller force can provide the same effects as massing force. This is particularly advantageous in the face of declining defense budgets, allowing commanders to get more from less.

NCW is derived from a concept from the business community, the most notable example of which is Wal-Mart's network-centric retailing. In Wal-Mart's platform-centric days, it had a central purchasing department that relied on inventory status reports from its various stores. This was replaced with point-of-sale scanners and inventory sensors that order replacements directly from the vendor as needed. The automated system also produces sales statistics on thousands of items, in real time, and marketing decisions are made accordingly, in near real time. Information on inventory, deliveries, and sales for any period of time is available to each store manager. The Navy and the C^{4}ISR community are pursuing the application of this concept to warfare (Cebrowski, 1998).

NCW employs the concept of three network grids:

1. **The information network grid:** The information grid provides the infrastructure to "receive, process, transport, store, and protect information for the Joint and combined services" (Stein, 1998). This is therefore the network-centric parallel to the information domain.

2. **The sensor network grid:** The sensor grid is a need-based network that makes use of the sensors in the information grid that are pertinent to a given task. It consists of such typical warfare sensors as

radar and imbedded logistics sensors to track supply. The sensor grid is unique to each task.

3. **The engagement-decision-shooter:** Sensor and warfighter elements of the network are tasked to attack in the engagement grid. This grid, like the sensor grid, is dynamic, using a unique blend of warfighters and sensors for each new task (Stein, 1998).

The second and third grids are completely contained in the information network grid. Each grid is composed of nodes represented by individual sensors, weapons, or command platforms and is connected via networked data and communications. The sensor and engagement grids are not necessarily separate, often having overlapping components. For example, the sensor grid begins a track on a cruise missile and continues to track as the pertinent unit engages and a kill is made.

NCW flattens the command and control pyramid. Commanders communicate intent by introducing doctrine in the form of computer algorithms and by communicating directly with individual units. NCW moves toward automated optimization of the positions of units in a group and engagement of enemy forces using new initiatives, such as the Navy's Cooperative Engagement Capability (CEC)[6] and Ring of Fire.[7]

Although NCW is not about networks exclusively, communications networks play a major role in achieving its goals. In a recently published book on the subject, Alberts, Garstka, and Stein (1999, p. 12) state that Network Centric Warfare is about

> exploiting information to maximize combat power by bringing more of our available information and warfighting assets to bear more effectively and efficiently. [It is] about developing collabora-

---

[6]CEC is designed to combine the raw sensor data from all platforms involved in an operation, regardless of age or type of sensors on individual platforms. It allows the combined data from these sources to produce a more complete, shared COP for tracking purposes. For additional information, see "The Cooperative Engagement Capability" (1995).

[7]The Ring of Fire (ROF) concept is a network-centric approach to littoral warfare. It links land, sea, and air forces to produce calls for fire. Like the CEC, ROF networks sensor and weapon information for sea, shore, and command forces in the littoral to produce an extended and more accurate COP (see Mitchell, 1998).

tive working environments for commanders, and indeed for all our soldiers, sailors, marines and airmen to make it easier to develop common perceptions of the situation and achieve (self-) coordinated responses to situations.

Exploiting information requires the ability to distribute fused information rapidly to all participating command nodes. This implies a robust communications network, as does the requirement for a collaborative working environment. One of the benefits of NCW is increased shared situational awareness of the battlespace.

COGNITIVE DOMAIN TERMS

The terms *situational awareness*, *shared situational awareness*, *cognition*, *understanding*, and *collaboration* appear frequently in discussions of combat decisionmaking. We defined *understanding* and *situational awareness* above, under the information domain, where they are also important concepts. Formal definitions of the remaining terms exist but are not always consistent and are sometimes not precise enough to satisfy the requirements of rigorous mathematical analysis. We tailored the following specific definitions to meet the requirements of this research:

- *Shared situational awareness*: the ability of a decisionmaking team to share their realization of the current situation. Note that based on this definition, *consensus* among the team members implies shared situational awareness.[8]

- *Collaboration*: a process in which two or more people work together to achieve a common objective. Collaboration requires the ability to share information.

- *Cognition*: the ability of humans to derive certain *knowledge* from an information source, such as a CROP. Note that cognizance does not imply either understanding or awareness.

---

[8]The Army's Digitization Office defines shared situational awareness as "the ability of a unit to know where its friends are located, where the enemy is, and to share that information with other friends, both horizontally and vertically, in near real-time." Again, this is a state-focused definition. It describes the state required to achieve shared situational awareness (U.S. Army Digitization Office, 1994).

However, the reverse is true, i.e., both awareness and understanding imply cognizance.

- *Common ground*: the knowledge, beliefs, and suppositions that individual collaborating team members believe they share. *Suppositions* are the propositions whose truths a team member takes for granted.

- *Belief*: a proposition that a participant would assent to if given ample opportunity to reflect. Along with facts and assumptions, beliefs constitute the foundation of common ground.

CANDIDATE MODELS

> Few things are harder to put up with than the annoyance of a good example.
>
> —*Mark Twain*

In the main text, we mentioned specific mathematical constructs that might be used to instantiate the general framework. This appendix offers a few of these, in addition to some more general methods mentioned in the text. It is important to understand that these expressions are purely speculative and are not the results of any in-depth research. However, they do illustrate the fact that reasonable mathematical expressions can be developed to measure the quality of information throughout the information superiority value chain.

SIMILARITY MATRICES

Similarity matrices serve to quantify generally nonquantifiable features by examining the "closeness" of the attributes. For example, suppose unit type is the feature to be assessed for a land force and that the complete set of possible attributes of unit type is $U =$ [Armored, Artillery, Mechanized, Reconnaissance]. Further assume that one way to characterize each type is by the equipment it usually possesses. For example, suppose that the complete set of major equipment types that characterizes each of these units is tanks, armored fighting vehicles, armored personnel carriers, and wheeled vehicles. Table B.1 offers a notional distribution of equipment for each of these unit types in terms of the fraction of the total equipment in the unit.

A table, such as Table B.1, can be transformed into a similarity matrix. *Similarity* or *dissimilarity* may be defined in terms of a distance metric, such as the Euclidean distance. The following example demonstrates how Table B.1 might be transformed into a similarity matrix.

Suppose we wish to calculate the "distance" between an armored unit and a mechanized unit. Using the Euclidean distance, we first calculate the "dissimilarity" as[1]:

$$D_{armor,mech} = \sqrt{\left(0.23 - 0.10\right)^2 + \left(0.19 - 0.30\right)^2 + \left(0.38 - 0.30\right)^2 + \left(0.19 - 0.30\right)^2}$$

$$= 0.218 .$$

Table B.1

Distribution of Equipment by Unit Type

	Tanks (%)	Armored Fighting Vehicles (%)	Armored Personnel Carriers (%)	Wheeled Vehicles (%)
Armor	0.231	0.192	0.385	0.192
Artillery	0.000	0.000	0.000	1.000
Mechanized	0.100	0.300	0.300	0.300
Reconnaissance	0.000	0.000	0.250	0.750

---

[1]Note that the maximum distance between any two units is the square root of two. This derives from the fact that the fractional equipment counts must sum to 1.0. The extreme case is two units, each consisting of only one type of equipment, which is different for each unit. For example, the distance between two units, one consisting only of tanks and the other only of armored personnel carriers, would be

$$D_{a,b} = \sqrt{\left(1.0 - 0.0\right)^2 + \left(0.0 - 1.0\right)^2 + \left(0.0 - 0.0\right)^2 + \left(0.0 - 0.0\right)^2} = \sqrt{2} .$$

The degree to which the two unit types are similar is then the complement of this figure: $S_{armor,mech} = 1 - D_{armor,mech} = 0.782$. This value is relatively high because armor and mechanized units are rather similar in composition of tracked vehicles. If we do this for each pair of unit types, the resulting nonnormalized similarity matrix will be as shown in Table B.2.

We next must assess the probability that a unit is correctly classified. That is, what is the probability that a unit of type i is not mistakenly classified as a unit of type j? To assess this likelihood, we first normalize the rows of the matrix in Table B.2 to get the matrix shown in Table B.3. The entries in Table B.3 can now be thought of as conditional probabilities. For example, the probability that an armored unit is correctly classified, given a report that the unit is mechanized, is $P(\text{armor}|\text{mechanized}) = 0.357$.

Suppose now that we interpret the entries as correctness values for estimates. That is, the correctness of an estimate that the unit is

Table B.2

Similarity Matrix for Unit Type

	Armor	Artillery	Mechanized	Reconnaissance
Armor	1.000	0.056	0.782	0.352
Artillery	0.056	1.000	0.175	1.646
Mechanized	0.782	0.175	1.000	0.448
Reconnaissance	0.352	0.646	0.448	1.000

NOTE: The diagonal elements are all 1.000 because all units are completely similar to themselves.

Table B.3

Normalized Similarity Matrix for Unit Type

	Armor	Artillery	Mechanized	Reconnaissance
Armor	0.4564	0.0257	0.3570	0.1608
Artillery	0.0300	0.5324	0.0934	0.3442
Mechanized	0.3252	0.0729	0.4158	0.1861
Reconnaissance	0.1440	0.2642	0.1830	0.4087

mechanized, given that it is really armored, is C(mechanized) = 0.357. In practice, we arrive at this estimate by observing n unit type reports for a single unit. If the most frequently reported unit type is mechanized, this is the estimate we use. The correctness of the estimate, however, is 0.357. Note that this correctness estimate is based on the similarity of a mechanized unit to an armored unit and that it presupposes that actual unit type is known. If the unit type is not known, we can assess the correctness of the report by assuming that the unit is indeed mechanized. In this case, we get C(mechanized) = 0.416.

The same matrix can be used to calculate the variance of the correctness of a sample of reported detections. Suppose we receive a sample of four reports of unit type U = {armor, mechanized, mechanized, artillery} .

The unit type estimate from this set of reports is "mechanized." If the unit is in fact an armored unit, as assumed above, the correctness of the estimate is U = C(mechanized) = 0.357, and the variance in the correctness is

$$V(U) = \frac{\sum_{i=1}^{4}(u_i - \overline{u})^2}{3}$$

$$= \frac{(0.456 - 0.357)^2 + (0.357 - 0.357)^2 + (0.357 - 0.357)^2 + (0.026 - 0.357)^2}{3}$$

$$= 0.0398 .$$

In general, for n reports, the variance is

$$V(U) = \frac{\sum_{i=1}^{n}(u_i - \overline{u})^2}{n-1} .$$

SET-THEORETIC TRACKING METRIC

If we assume that the number of active tracks at time $t - 1$ is X_{t-1} and that at scan t it is X_t, the question about how well is the sensor system tracking targets depends on the relative size of X_{t-1} and X_t. In the main report, we discussed the possibilities using some assumptions.

Here, we wish to expand a bit more on the development of the suggested tracking metric.

Figure B.1 depicts a generalized representation of the two sets of tracks. To develop a metric, $0 \leq T \leq 1$, we examined the size of the intersection, $A \cap B$. Tracks from scan $t - 1$ that are still active in scan t reside in the intersection. The number of tracks in $A \cap B$ is $|X_{t-1} - X_t|$. An appropriate metric therefore might be

$$T = \frac{\left|X_{t-1} - X_t\right|}{X_{t-1}}.$$

This metric satisfies the requirements established in the main report that, when $A \cap B = \varnothing$ (*no* previous tracks are active), $T = 0$; when $A \cap B = A$ (*all* previous tracks are active), $T = 1$; for all other possibilities, $0 < T < 1$.

CRITICAL PATH METHOD

The CPM is used to assess the time required to complete a process (see, for example, Taha, 1976). This section illustrates this methodology. This project management tool was designed to identify tasks in a project that require special attention because slippage in these will cause slippage in the completion of the project. The critical path defines a chain of critical activities that connect the start and end of events in a directed network, such as that depicted in Figure B.2. It reflects the time required to traverse the longest (most time consuming) path in the network of tasks.

RAND *MR1467-B.1*

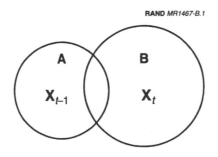

Figure B.1—Track Sets

RAND *MR1467-B.2*

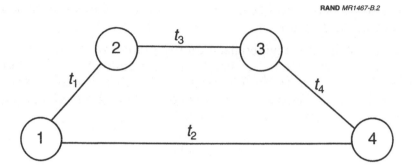

Figure B.2—Example Currency Calculation

In assessing the currency of information, the central issue is which process times recorded are sequential and which are parallel. For example, sensor reports cannot be combined to produce the CROP until the they have been collected. The total time elapsed in the information domain, then, is either the sum of the times required in each subdomain, the maximum time across subdomains, or some combination of both. Although the actual calculation will depend on the architecture designed for each given alternative, we can illustrate the process with an example.

Suppose we assign times to the various tasks defined for each sub-domain. We have not attempted to decompose these tasks into sub-tasks, although in practice, this is highly recommended. Figure B.2 graphically represents the sequencing for four tasks. The circles represent events, and the arcs represent processes to complete in the time specified. The structure of the graph is situation dependent, as the discussion of the events and the tasks will make clear.

Event 1. This starts the sensing process. It can be considered the initial tasking to the sensor suite. Two tasks begin at once: Initial detection and target tracking and retasking of sensors to capture high-value targets—if that capability exists.

Event 2: Detection and tracking are complete. A better interpretation might be that sufficient sensor data are available to allow the fusion process to begin. This is represented by task 3.

Event 3: A CROP is produced. This cannot occur until the development of the CROP is complete (task 3). Note that it does not depend on the retasking. The assumption for this example is that retasking continues and that updates are processed as they arrive or at specified intervals.

Event 4: This is actually two events. The first is the dissemination of the CROP to the users, and the second is the completion of retasking. At this point, the CROP is complete and has been disseminated to the users.

The composite system currency, given these assumptions concerning the processes, obtained by inspection, is then

$$t = \max\{(t_1 + t_3 + t_4), t_2\}.$$

A more systematic approach for more complex task and event groupings can be found in any standard operation research text.

SENSORS

Figure B.3 illustrates a typical model of sensor *detection* performance as a function of *range*. A sensor with an unobstructed view of the battlespace will generally have a minimum and a maximum range. Within this range band, the sensor is capable of detecting targets of a type specific to its technical capabilities. The maximum range will vary according to several factors. For example, the available power and, if the sensor is used to classify targets, its resolution will affect an active sensor's maximum range. The minimum range may be zero or may be a significant distance, as for radar and imaging systems.

Within its operational range, a sensor will have a nonzero probability of detecting its designated targets. In Figure B.3, we have the following functional relationship:

$$P(d) = \begin{cases} k & \text{if } d_{\min} \leq d < d_b \\ f(d) & \text{if } d_b \leq d \leq d_{\max} \\ 0 & \text{otherwise}. \end{cases}$$

The sensor achieves its maximum capability, $P(d) = k \leq 1$, at the minimum range and maintains that capability through range d_b.

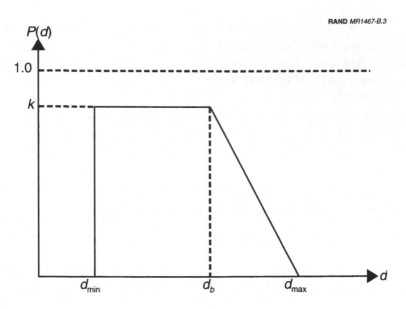

Figure B.3—Generic Sensor Performance Model

Performance begins to fall off beyond d_b according to some relationship (depicted as a straight line in the diagram). The functional relationship may vary as a function of a number of factors, including target aspect angle relative to the sensor, target state (temperature, velocity, configuration, etc.), and the state of the immediate and intervening environments. The generic profile in Figure B.3 can be used to model the performance of many sensor types.

A Reliability Model of Completeness

It seems natural to equate the probability that a sensor will detect a target with its reliability. Because the completeness of the reports from a single sensor is a function related to its reliability, we might choose sensor reliability as a metric for the completeness of the sensor reports. The problem is then to examine the nature of $P(d)$ for each sensor and for the entire suite. One way to do this is to assess the reliability of the sensor and the sensor suite. We let $P_i(d) = R_i(d)$ be the reliability of sensor S_i, where d is the distance from the sensor to the target. Generally, the independent variable is time, but it need not be. In the case of sensors, as discussed above, distance is much more useful in that it can account for malpositioned sensors and inefficient multisensor configurations and because it generally characterizes sensor detection performance. The general form of $R(d)$ is

$$R(d) = e^{-\int_0^d r(s)ds} ,$$

where $r(d)$ is the failure rate function and is dependent on the characteristics of the sensor and the current operating situation. Sensor characteristics include the ability to detect, estimate, and classify targets. Note that, for $d = 0$, $R(d) = 1$. That is, when the sensor is coterminal with the target, it is infallible. This is, of course, an idealization; therefore, to conform more closely to the generic model, we recast the reliability function as the decreasing segment in a piecewise relationship so that we now have

$$R(d) = \begin{cases} 0 & \text{if } d \leq d_{\min} \\ e^{-\int_{d_{\min}}^d r(s)ds} & \text{otherwise.} \end{cases}$$

Note also that $R(d)$ is a probability that can be interpreted as the probability of detection at range d.[2] In this formulation, $d_{\max}$ is reached only in the limit. This is not much of a problem, however, since the rate of decline of $R(d)$ for $d > d_{\min}$ is controlled by $r(d)$. Depiction of the linear decline is more problematic. In this case, the reliability formulation should be replaced with

$$R(d) = \begin{cases} 0 & \text{if } d \leq d_{\min} \\ \dfrac{d_{\max-d}}{d_{\max} - d_{\min}} & \text{otherwise.} \end{cases}$$

Expressing sensor detection probability in this way can be useful. For example, the effects of occlusions can be modeled through the appropriate selection of $r(d)$ so that for a totally occluded view, $d \to \infty$; for impaired views, such as foliage, atmospheric disturbances, etc., d is set to be larger than the physical distance between the sen-

---

[2]There are several good texts on reliability engineering. See, for example, Ayyub and McCuen (1997) and Pecht (1995).

sor and the target. This suggests that $r(d)$ might be defined piecewise to reflect the effects of occluded views. We pursue this more fully below.

As an example, suppose we have three sensors whose detection probabilities beyond the minimum range, d_{min}, are determined by the following failure rates: $r_1(s) = 1$, $r_2(s) = s$, and $r_3(s) = s^2$. This leads to the following detection probability functions:

$$R_1(d) = e^{-(d - d_{min})},$$

$$R_2(d) = e^{-\frac{1}{2}(d^2 - d_{min}^2)}, \text{ and}$$

$$R_3(d) = e^{-\frac{1}{3}(d^3 - d_{min}^3)}.$$

Figure B.4 depicts varying detection probabilities for these three sensors, with $d_{min} = 0$ for simplicity. Note that the relative behavior of the sensors, as well as their absolute behavior, is dependent on their distance to the target. Sensor 3, for example, is preferable when the target is close in, but sensor 1 outperforms the other two when the target is at a greater distance. From an operational point of view, therefore, it is important to know where the intersection of the performance lines is, as depicted in the diagram. This is equivalent to knowing the relative performance of the sensors, within their specified performance bands.

However compelling this construct might be, it is important to note that it ignores the importance of false targets. For example, Sensor 3 may be preferable up to 1.5 km in Figure B.4, but that may also mean that is more likely to detect false targets at that range if we assume that a sensor is as likely to detect false as it is to detect real targets. The problem is one of discrimination. How well can the sensor differentiate real from false targets and at what ranges?

Occlusions

A sensor is occluded when terrain and/or foliage intervene between the sensor and the target. Most sensors require a clear, unobstructed

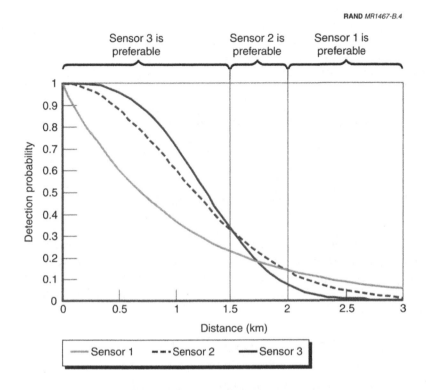

Figure B.4—Sensor Performance

view of the target. Targets can be hidden from view by the terrain features, for example mountain ranges. The elevation profile of the terrain in the AO can be considered to be a geometric surface, $z = g(x, y)$, where x and y are planer coordinates, such as latitude and longitude or Universal Transverse Mercator (UTM) coordinates. Figure B.5 illustrates a typical AO, in which a sensor is located at some elevation at (x_0, y_0), and a target is located on the surface at (x_1, y_1). The area within which ground targets can be detected is indicated in medium gray. The area within which ground targets cannot be detected is indicated in light gray. The sensor can detect all the targets within the valley in the foreground but none of those beyond the mountain range (for example, those in the next valley). Complex geometric relationships can exist between targets, sensors, and the environment.

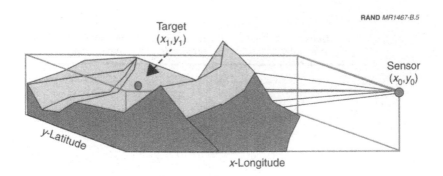

RAND *MR1467-B.5*

Figure B.5—Terrain Occluded Targets

We can also use the reliability function to model the probability of detection for different levels of occlusion without resorting to complex surface maps. For example, suppose the failure rate for a given sensor without occlusions is $r(s) = 1$. This produces a detection probability function like that for Sensor 1 above:

$$R(d) = e^{-(d - d_{min})}.$$

We can now express the effects of occlusions by damping $R(d)$, so that

$$R(d) = ke^{-d - d_{min}},$$

where $k \in [0,1]$. For $k = 0$, we have a totally occluded sensor, and, as mentioned above, $R(d) = 0$ at all distances. For $k = 1$, no occlusions exist, and we get the basic relationship. All other values of k reflect varying levels of occlusion, and their effect is to reduce the probability of detection. Figure B.6 depicts the extreme cases and two that are intermediate. We assume, as in Figure B.4, that $d_{min} = 0$.

Another use of $r(d)$ might be to assess the effectiveness of sensor tasking. When sensors are tasked to focus on a particular aspect of the battlespace, the practical effect is to reduce the distance between the sensor and the target—thus improving the reliability of the tasked sensors. For example, Sensor 1 in Figure B.4 has a probability

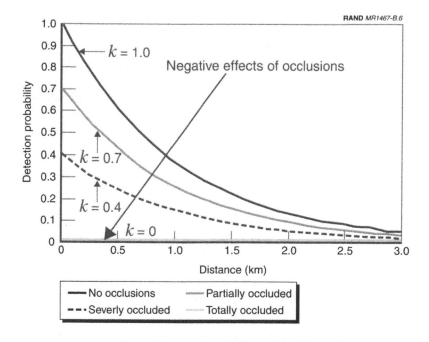

Figure B.6—Accounting for Occlusions

of detection of 0.368 at 1 km from the target, so $R(1) = 0.368$. Suppose this is the closest the sensor is able to approach the target, and further suppose that only Sensor 1 types are available. If we task three of these to observe the target, the detection probability becomes

$$R_3(d) = 1 - (1 - 0.368)^3.$$

For a Sensor 1 type, this is equivalent to being approximately 0.3 km from the target; therefore, $R_3(0.3) \approx 0.748$.

Collective Completeness

The better the sensor suite (in terms of sensor performance and operational integration), the more likely it is that the number of targets detected will be the total in the AO. Given an expression for the reliability of individual sensors, what is needed is an assessment of

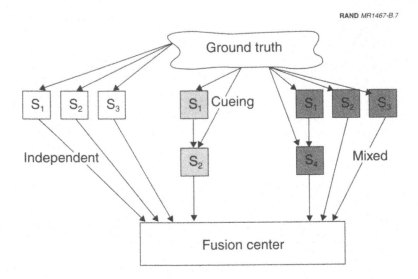

Figure B.7—Multisensor Operations

the integrated sensor suite. Figure B.7 depicts three possible operational modes: independent operation, cueing, and mixed mode.[3] The structure of the network is exemplary. The fact that all reports end at a single fusion center is not central to the assessment. Clearly, fusion can take place anywhere in the ISR and fusion system and may be either distributed or centralized. Indeed, one value of measuring the quality of information generated from sensor operations and fusion processes is to evaluate the effectiveness of alternative architectures.

It is important to note here that these alternative architectures are not to be confused with alternative multisensor decision structures (see Waltz and Llinas, 1990). A multisensor decision structure addresses the way complementary sensors arrive at a detection decision before transmitting the decision to the fusion center. In a centralized decision structure, each sensor passes its observations directly to an observation facility within the fusion center, where a combined decision is taken. In a decentralized structure, each sensor

---

[3]This may not be an exhaustive set. For example, a standby configuration is also possible, in which a less-reliable sensor is held in reserve to temporarily replace a more-reliable system that has failed. See, for example, Dhillon (1988).

renders a decision, which is forwarded to the fusion center. The architecture in Figure B.7 can support either structure.

Independent Operation. This is the simplest form of operation. Each sensor observes the AO independently and reports its detections to the common fusion center. This is essentially parallel operation in reliability analysis, and it increases the likelihood that a detection(s) will occur. To calculate the sensor suite reliability for this case, we used

$$R(d) = 1 - \left[1 - R_1(d)\right]\left[1 - R_2(d)\right]\left[1 - R_3(d)\right].$$

Cueing. In cueing operations, one sensor detects a target and notifies another to confirm the detection (or, in practice, to provide additional data on the target). A report is rendered when the two sensors have detected the target. This results in a conditional reliability, $R_{1|2}(d)$, that is, the reliability of sensor S_1 given the reliability of sensor S_2. In this formulation, the two sensors are not independent. That is, the reliability of the cued sensor depends on the reliability of the cueing sensor. This further requires that a failure rate, $r_{1|2}(s)$, be determined for each cued-cueing pair.

Mixed Mode. This is the most likely operational mode, a mixture of both independent operations and cueing. The overall system reliability in this case is dependent on the complexity of the system structure. For the simple case depicted in the diagram, this is

$$R(d) = 1 - \left[1 - R_1(d)R_4(d)\right]\left[1 - R_2(d)\right]\left[1 - R_3(d)\right].$$

A reasonable estimate for the number of targets in the AO detected by the sensor suite at time t then is $100R(d)$. If, for example, 10 targets are detected, and $R(d) = 0.4$, the estimated total number of units or targets in the AO is $10/0.4 = 25$. The poor reliability of the sensor system in this example is a function both of the system architecture and the distances the sensors are to the targets.

Assuming the fraction of the AO the sensor suite covers is c_2, the following estimates the completeness transformation metric using the suggested reliability model:

$$Q_{com}\left(F_1 \mid F_0\right) = R(d)\left(1 - e^{-c_2}\right).$$

DISSEMINATION

In Chapter Three, we suggested that one way to assess the completeness of the dissemination process is to assess the reliability of the
dissemination network. This, in turn, requires that we assess the reliability of the links in the network. In this section of the appendix, we
illustrate the process by postulating a small subnetwork consisting of
a single fusion center, F; three CROP users, U_1, U_2, and U_3; and two
relay nodes, T_1 and T_2 (transshipment nodes, in the language of network theory). Figure B.8 illustrates the connectivity among the
nodes. We assumed that the network is cyclic—that two-way communication is possible on all links. However, we ruled out cycles in
communicating between nodes and assessed the likelihood that the
three user nodes are connected. Clearly, this depends on the reliability of the links on the various paths from the fusion center to the user
nodes.

Using a reliability model, we focus on link (L_i) reliability, $R_i(q)$, where
the failure rate function, $r(q)$, is a measure of communications quality and, as such, is a function of the SNR, jamming, bandwidth, etc.

RAND *MR1467-B.8*

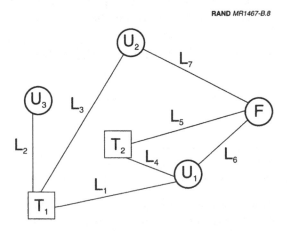

Figure B.8—A Communications Subnetwork

Depending on the convention adopted, we can have $R_i(q)$ increase as q increases or the opposite. For example, if $r(q) = q$,

$$R_i\left(q\right) = e^{-\frac{q^2}{2}},$$

and $R_i(q)$ decreases with increasing q. This can be used to model the effects of jamming, for example. However, in the case of SNR and bandwidth, we would expect the reliability of the link to increase with increasing values of q. To get this opposite relationship, we can set

$$R_i\left(q\right) = 1 - e^{-\frac{q^2}{2}}.$$

The probability that a user is connected is then the probability that at least one path between the fusion facility and the user is available, given the values of q for each of the links. The information transmitted over the network is then the joint probability that all the users are connected at time t. For the simple network in Figure B.8, we completed the calculations depicted in Table B.4. The last column is the probability that the individual user is connected; therefore, the

Table B.4

Network Completeness Assessment

User	Path	Path Reliability	Probability Connected, P_i
U_1	L_6	$R_6(q)$	$1 - [1 - R_6] [1 - R_5 R_4]$
	$L_5 \rightarrow L_4$	$R_5(q) R_4(q)$	$[1 - R_7 R_3 R_1]$
	$L_7 \rightarrow L_3 \rightarrow L_1$	$R_7(q) R_3(q) R_1(q)$	
U_2	L_7	$R_7(q)$	$1 - [1 - R_7] [1 - R_6 R_1 R_3]$
	$L_6 \rightarrow L_1 \rightarrow L_3$	$R_6(q) R_1(q) R_3(q)$	$[1 - R_5 R_4 R_1 R_3]$
	$L_5 \rightarrow L_4 \rightarrow L_1 \rightarrow L_3$	$R_5(q) R_4(q) R_1(q) R_3(q)$	
U_3	$L_6 \rightarrow L_1 \rightarrow L_2$	$R_6(q) R_1(q) R_3(q)$	$1 - [1 - R_6 R_1 R_2] [1 - R_7 R_3 R_2]$
	$L_7 \rightarrow L_3 \rightarrow L_2$	$R_7(q) R_3(q) R_2(q)$	$[1 - R_5 R_6 R_1 R_2]$
	$L_5 \rightarrow L_6 \rightarrow L_1 \rightarrow L_2$	$R_5(q) R_6(q) R_1(q) R_2(q)$	

assessment of complete network connectivity is the probability that all users are connected:

$$N = \prod_{i=1}^{3} p_i \, .$$

INDIVIDUAL SITUATIONAL AWARENESS

In Chapter Four, we discussed the relationship between the awareness of the alternative decision agents and each of the total information quality measures. We used $A \in [0,1]$ to denote individual awareness. This section illustrates two possible functions relating information completeness to the awareness that decision agents with varying capabilities, denoted Φ_i, can achieve. For both functions, we generated a set of isocurves that vary with Φ_i. Accordingly, we modified the awareness metric to A_{Φ_i}.

The first is a series of exponential functions of the form $A_{\Phi_i} = 1 - e^{-\beta C}$. The shape of the curve depends on the parameter β; the parameter's value is chosen to reflect the decision agent's characterization. For example, the curve is shifted to the right to reflect the "unaware" decision agent's (Φ_4) requirement for high degrees of completeness. Figure B.9 illustrates the four curves discussed in the text.[4]

The so-called logistic or S-curve may also describe a decision agent's variable behavior.[5] That is, for low levels of information quality, his awareness is at its lowest level. For some region of awareness above this threshold, his awareness increases rapidly, tapering off considerably beyond this region. Figure B.10 illustrates this behavior. These curves are all of the form

$$A_{\Phi_i} = \frac{e^{\beta_0 + \beta_1(\Phi_i)C}}{1 + e^{\beta_0 + \beta_1(\Phi_i)C}} \, . \qquad \text{B.1}$$

---

[4]This figure also appears in the main text as Figure 4.2.

[5]This curve is sometimes referred to as the *logistics response function* or the *growth curve* (see Neter and Wasserman, 1974).

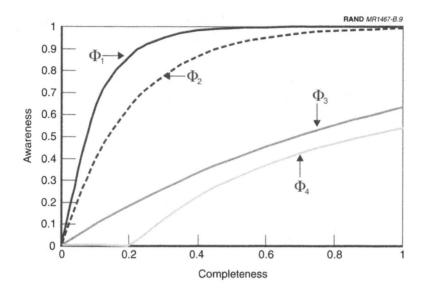

**Figure B.9—The Effect of Complete Information
on Situational Awareness**

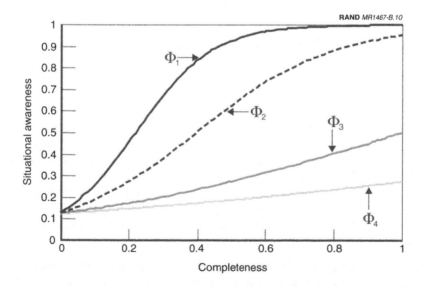

**Figure B.10—S-Curve Representation of Completeness
on Situational Awareness**

The parameters β_0 and $\beta(\Phi_i)$ reflect the decision agent's characterization and may be determined through analysis of variance techniques.

The relationship defined in equation B.1 is more generally applicable to quality of information. Extending equation B.1, the general relationship, yields

$$A_{\Phi_i} = \frac{e^{\beta_0+\beta_1(\Phi_i)Q(F_0,F_1,F_2,F_3)}}{1+e^{\beta_0+\beta_1(\Phi_i)Q(F_0,F_1,F_2,F_3)}} \,,$$
B.2

where $Q(F_0,F_1,F_2,F_3)$ is the quality of the observed CROP.

In both formulations, A_{Φ_i} is interpreted to be the degree of individual situational awareness—the fraction of the observed CROP that an individual decisionmaker with ability Φ_i is able to realize. It is parametrically related to Φ_i, in that we postulate a finite set of capabilities $\beta_1(\Phi_i)$ and adjust accordingly.

From Figure B.9, the degree of individual situational awareness that the four decision agent capability levels (Φ_1, Φ_2, Φ_3, and Φ_4) attain for a CROP with quality $Q(F_0,F_1,F_2,F_3) = 0.4$ is approximately 100, 88, 35, and 25 percent of the fused feature vectors in the CROP, respectively. For Figure B.10, it is 85, 52, 25, and 17 percent of the fused feature vectors, respectively.

Transactive Memory System Model

A transactive memory system, A, is a set of individual memory systems in combination with the communication that takes place between individuals. The transactive memory system model addresses the storage and movement of information among actors. In Chapter Four, we discussed a general construct for measuring what we referred to as "team hardness" in such a system. This section suggests a specific model that possesses the appropriate characteristics.

A basic assumption of the model is that the rate at which the ratio of shared information storage or complexity in the transactive memory grows is linear. The rate depends on the degree of consistency of team membership over time and on the complexity of the situation.

We represented the rate at which hardness increases as a parameter, k, expressed in units of hardness per unit time. With a constant growth rate model, a simple increasing exponential can represent team hardness. If we let t represent the time elapsed since the start of the operation and τ the length of time the team has been training or operating together, transactive memory is $TM(T) = 1 - e^{-kT}$, where $T = \tau + t$. The operation is assumed to have begun at time $t = 0$. Figure B.11 depicts the behavior of $TM(T)$ for various values of k.

Teams with high team hardness rates (k_1) rapidly develop the shared information and transactive memory system complexity necessary to respond to a given situation. Teams with moderate team hardness rates (k_3) must refine a transactive memory system while responding to the current situation. Teams with a low team hardness rate (k_4) must establish and develop the transactive memory system while responding to the current situation.

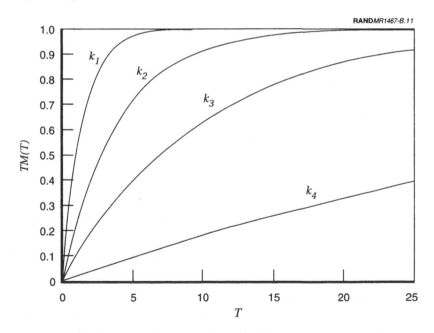

Figure B.11—Transactive Memory for Alternative
Values of Team Hardness

A Revised Model of Situational Awareness

The next logical step is a revised measure of individual situational awareness that combines the quality of the observed CROP and the effects of the transactive memory system, $A_{\Phi_i}(t)$. This measure is time dependent because of the introduction of transactive memory. It is now interpreted to be the fraction of the observed feature vectors that the individual decisionmaker with capability Φ_i and with benefit of team participation can realize when the operation has been in progress for t time units.

$$A'_{\Phi_i}(t) = \frac{e^{\beta_0+\beta_1(\Phi_i)\left[Q\left(F_0,F_1,F_2,F_3\right)+TM(T)\right]}}{1+e^{\beta_0+\beta_1(\Phi_i)\left[Q\left(F_0,F_1,F_2,F_3\right)+TM(T)\right]}} .$$

Note that if $\beta \to \infty$ at time $t = 0$,[6] $Q[O(t)] = 0$ and $TM(T) = TM(\tau) = 0$ for a team with no collective experience.

A MODEL OF SHARED SITUATIONAL AWARENESS

As Chapter Four suggested, the next step is to evaluate the situational awareness of the team when it is working together, $SA(t)$, recognizing that this measure will also will be time dependent. This was referred to as *shared situational awareness*. The general relationship is expressed as

$$SA(t) = g\left(A'_{\Phi}, F_5\right).$$

the collective fraction of the observed CROP, $Q(F_0,F_1,F_2,F_3)$, that the entire team can interpret. In this formulation, F_5 represents the observed CROP agents of capability Φ_i at time t can realize. What follows is a suggested methodology for calculating $SA(t)$.

Suppose we let m be the number of feature vectors in the observed CROP, i.e., the cardinality of the set F_5 is m, or $\|F_5\| = m$, and the number of feature vectors realized by all team members with capability Φ_i at time t is

---

[6]In practice, a value of -5 is sufficient.

$$mA'_{\Phi_i}(t).$$

However, the feature vectors interpreted may not be the same for each individual. For team member j with capability Φ_i, the cardinality of the set of feature vectors an individual can interpret is therefore

$$\left\| \mathbf{F}_{j\Phi_i} \right\| = mA'_{\Phi_i}(t).$$

Since it is impossible to know which vectors are realized, we instead examined the possible overlaps. The intersection set between two team members, j and k, one with capability Φ_i, and one with capability Φ_l is given by

$$\mathbf{F}_{j\Phi_i} \cap \mathbf{F}_{k\Phi_l}.$$

The smallest number of elements in this set (overlap) is

$$\left| A'_{\Phi_i}(t) - A'_{\Phi_k}(t) \right| m,$$

and the largest it can be is

$$\min\left\{ mA'_{\Phi_i}(t), mA'_{\Phi_k}(t) \right\}.$$

Therefore, a reasonable estimate of the fraction of overlapping feature vectors that team members j and k with capabilities Φ_i and Φ_l realize is the average of these two quantities:

$$\begin{aligned} G_{jk} &= \frac{1}{2}\left[\left| A'_{\Phi_i}(t) - A'_{\Phi k}(t) \right| + \min\left\{ A'_{\Phi_i}(t), A'_{\Phi k}(t) \right\} \right] \\ &= \frac{1}{2}\max\left\{ A'_{\Phi_i}(t), A'_{\Phi_k}(t) \right\}. \end{aligned}$$

B.3

For example, if $i = l$,

$$G_{ik}(t) = (1/2)A'_{\Phi_i}(t).$$

This appears to be right, in that, on average, two team members that realize the same number of feature vectors will, on average, have half in common. Equation B.3 therefore represents the degree of consensus between two team members.

Next, we needed to account for the composition of the team itself, that is, for the number of each capability type present in the team and the size of the team. We might have done this by assessing the degree of consensus among all pairwise combinations of team members. We paired all possible capability types, making the calculation above and averaging the result, or, for a team of size n:

$$SA(t) = 1/\binom{n}{2} \sum_{i=1}^{n-1} \sum_{j=i+1}^{n} G_{ij}(t).$$
                                                                    B.4

The degree of shared situational awareness, $SA(t)$, is the collection of fused feature vectors at time t that members of a team realize similarly if they collaborate.

Equation B.4 approximates the degree of shared situational awareness because our measure for the degree of consensus is an average. We also assumed that all pairwise combinations of collaborating individuals have the average measure for degree of consensus. Despite these limitations, equation B.5 does tie together the individual and team components of shared situational awareness. We acknowledge that more-sophisticated relationships may be developed but believe that the equation captures the essential components necessary for further research into shared situational awareness.

SPREADSHEET MODEL

There is no study that is not capable of delighting us after a little application to it.

—Alexander Pope

We designed measures and metrics the main report describes and Appendix B details to reflect how the quality of the information produced within the three C^4ISR domains and during team collaboration affects shared situational awareness. The assumption throughout has been that the process that produces the information (sensors, fusion facilities and processes, distribution networks) is not necessarily known, and we are not concerned about how the information is transformed as it moves through the information value chain. The quality of the processes is expressed through the quality metrics in the form of parameters and alternative probability functions. The quality of team interactions is assessed by examining the capabilities of individual team members, the degree of team hardness, and the level of collaboration.

This appendix illustrates the assessment process through a simple spreadsheet model that allows us to alter quality parameters. The reliability curves are taken to be constant, except that the quality parameters can be varied.

MODELING INFORMATION QUALITY

A simple spreadsheet model was developed to illustrate the information quality assessment process.[1] Only the completeness measures are included in the version described below. However, the model can easily be expanded to include correctness and currency. Time, t; the distance between the target and the sensor, d; and the level of jamming, q, are independent variables. A three-dimensional response surface (the composite completeness) is displayed using time on one axis and either distance or the level of jamming on the other. The independent variable not used is treated as a parameter.

Sensors

We modeled three sensors, each with a unique reliability function (see Table C.1). In all cases, we took the minimum distance for detection to be 0. The occlusion settings, k_i, varied, as did the distance between the sensor and the target.

Table C.1

Sensor Quality Functions

Sensor	$r(s)$	$Ri(d)$
S_1	1	$k_1 e^{-d}$
S_2	s	$k_2 e^{-\frac{d^2}{2}}$
S_3	s_2	$k_3 e^{-\frac{d^3}{3}}$
$S_1 \mid S_3$	0.5	$k_4 e^{-0.5d}$
$S_1 \mid S_2, S_3$	0.333	$k_5 e^{-0.333d}$

---

[1]The model is titled *Infoview* and was designed and implemented by RAND colleague Thomas Sullivan.

We examined three architectures: independent operation, queuing, and a mixture of both. Figure C.1 describes all three. The system reliabilities for all three are as follows:

a.   independent operations:

$$R(d) = 1 - \left[1 - R_1(d)\right]\left[1 - R_2(d)\right]\left[1 - R_3(d)\right]$$

$$= 1 - \left[1 - k_1 e^{-d}\right]\left[1 - k_2 e^{-\frac{d^2}{2}}\right]\left[1 - k_3 e^{-\frac{d^3}{3}}\right].$$

b.   queued operations:

$$R(d) = R_{1|2,3}(d) = k_5 e^{-0.333d}.$$

c.   mixed operations:

$$R(d) = 1 - \left[1 - R_{1|3}(d)\right]\left[1 - R_2(d)\right] = 1 - \left[1 - k_4 e^{-0.5d}\right]\left[1 - k_2 e^{-\frac{d^2}{2}}\right].$$

If we further assumed that the sensors for all three architectures covered the same portion of the area, $c_2 = 0.8$, yielding the following completeness metric: $Q_{com}(F_1|F_0) = R(d)(1 - e^{-0.8})$.

Fusion

We modeled a single fusion facility consisting of IMINT and ELINT processing centers and a central processing center, as depicted in Figure C.2. Each of the three facilities was modeled using the metric suggested in Chapter Three:

$$c = a + \alpha\left(1 - e^{-\sigma t}\right).$$

Recall that a represents the fraction of the detected targets that the sensors themselves can fuse and that $0 < a + \alpha \le 1$ is the maximum fraction of the detections capable of being classified at the fusion center. The third parameter, σ, is the rate at which detections are

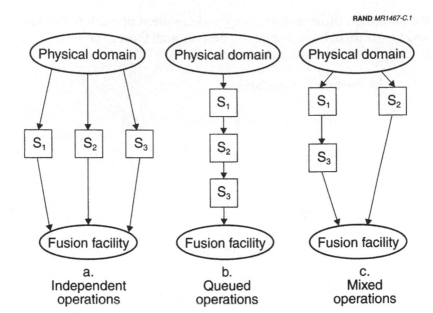

RAND *MR1467-C.1*

Figure C.1—Alternative Sensor Architectures

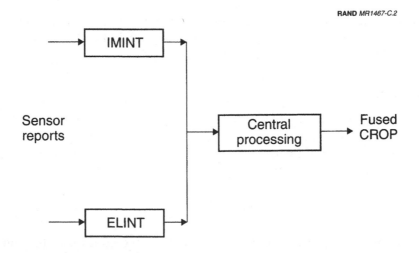

RAND *MR1467-C.2*

Figure C.2—Model Fusion Architecture

classified. These parameters are set in the model as shown in Figure C.3. From this the fusion architecture, we can calculate completeness as follows:

$$C = [1 - (1 - c_{IMINT})(1 - c_{ELINT})]\, c_c \,,$$

where cc represents the completeness of the central fusion facility.

Network

The network modeled is the sample network in Figure B.8. For all links in the network, we modeled the completeness of the messages transmitted in terms of the level of jamming. In addition, we assumed that the effects would be the same for all links and that they are defined by the failure rate function $r(q) = q$. Therefore, the completeness metric for each link is as follows:

$$R_L(q) = e^{-\frac{q^2}{2}}\,.$$

Applying this to the network results in a network completeness metric, N, calculated using the procedures presented in Table B.3.

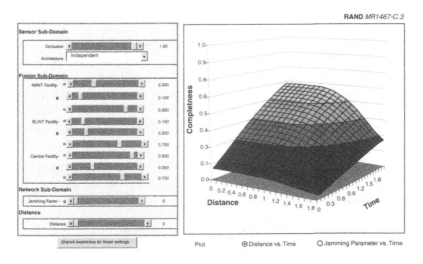

Figure C.3—Infoview Information Domain GUI

MODELING TEAM DECISIONMAKING

We expanded the spreadsheet model above to account for the next step in the process. The quality of information produced thus far becomes one factor in the overall assessment of situation awareness. Typically, the parameter settings established to produce a given level of information quality in the information domain are "carried over" into the cognitive domain.

The additional parameters needed to complete the assessment of shared situational awareness are the degree of team hardness and the number of team participants possessing the different levels of capability. The equation used to produce the level of transactive memory achieved depends on the parameter selected for team hardness. In this case we selected a value of $k = 1.0$, leading to

$$TM(T) = 1 - e^{1.0T}.$$

We must now assess the effect of the level of individual situational awareness that the members of the team with capability Φ_i possesses for each capability level. For this example, we assumed four levels of capability, and therefore

$$A_{\Phi_i'}(t) = \frac{e^{-5+\beta_1(\Phi_i)\left[Q(F_0,F_1,F_2,F_3)+TM(T)\right]}}{1+e^{-5+\beta_1(\Phi_i)\left[Q(F_0,F_1,F_2,F_3)+TM(T)\right]}}$$

represents the fraction of the observed feature vectors that the individual decisionmaker with capability Φ_i and with benefit of team participation can interpret. For this example, we assumed that there are nine team members: two with capability Φ_1, three with capability Φ_2, one with capability Φ_3, and three with capability Φ_4. Note also that we set $\beta_0 = -5$. The coefficient β_1 is referred to as the threshold and is set at 0.15 for this example.

It is not necessary that we know the number of feature vectors in the observed CROP, since we are only concerned with the average fraction of overlapping features:

$$G_{jk}(t) = \frac{1}{2}\max\left\{A_{\Phi_i}(t), A_{\Phi_i}'(t)\right\}$$

for every pair of participants in the decisionmaking team, j, k.

With this, the total shared situational awareness is calculated to be:

$$SA(t) = 1 / \binom{n}{2} \sum_{i=1}^{n-1} \sum_{j=i+1}^{n} G_{ij}(t).$$

THE GRAPHICAL USER INTERFACES

Two graphical user interfaces (GUIs) are used to assess shared situational awareness. The first depicts information domain effects on information quality, and the second depicts the effects of information quality and cognitive domain conditions on shared situational awareness.

Information Domain Assessments

Figure C.3 depicts the GUI for the completeness model only. The slide bars to the left of the figure allow the user to select the desired parameter settings. Beginning with the sensor subdomain, the user selects the level of occlusion and the sensor architecture. The two INT disciplines, IMINT and ELINT, and the central processing facility are displayed next, each with three parameter sliders. The first, σ, determines the rate at which the observations are classified; the second parameter, a, represents the fraction of the detected targets that can be fused by the sensors themselves; and the third parameter, α, is selected such that $a + \alpha$ is the upper bound on the fraction of detections that can be classified. Recall that combinations of automation and control protocols can affect the classification rate at each of the fusion facilities. The effects are therefore reflected in the values of the parameters a, α, and σ. The sole network parameter, q, is the level of jamming on all links. If the "Distance vs. Time" button is selected from the "Plot" options under the graph, q is defined parametrically, since d is the selected independent variable. The opposite is true if "Jamming Parameter vs. Time" is selected.

The three-dimensional figure to the right represents the degree of completeness for a given sensor distance and processing time. The decline along the distance axis for a given processing time reflects the fact that completeness deteriorates as the distance from the sensor to the target increases. The behavior along the time axis for a given distance reflects the fact that completeness is likely to increase with more time available to process information.

Finally, note the selection bar titled "Shared Awareness for these settings" at lower left. The effects of information quality and the other factors on shared awareness are evaluated when this button is selected. This button sets the completeness quality level that is input to the evaluation of shared awareness. Selecting this option will take the user to another GUI that will be discussed next.

The three-dimensional surface to the right in Figure C.3 results from the settings selected at the left. In this case, the maximum completeness is approximately 60 percent and occurs at approximately 2 time units, when the sensors are roughly between 0 and 1 distance unit from the target. Note that completeness falls off as the distance from the target increases. Note also that completeness rises rather rapidly with time. This is due to the shape of the fusion curves and thus reflects the degree of control and automation.

Cognitive Domain Assessments

Figure C.4 depicts the GUI for evaluating shared situational awareness. As with the information domain GUI, the slider bars at the left allow various parameter settings. The first bar sets the value of β_1 for all decision agents—0.15 for this example. The second bar establishes the value for team hardness (1.8 in this case). The next four bars set the number of decision agents at each of the four capability levels participating in the decision. The settings depicted are consistent with the numbers stated above. Note also that $\beta_0 = -5$, as suggested above.

The smaller box at the lower right records the values of the parameters set in the information domain GUI. These settings are used to calculate the information quality at the various times, t, during the decisionmaking process. Information domain settings are constant for all parameter settings in the cognitive domain GUI.

Finally, the three-dimensional figure depicted represents the degree of shared situational awareness, $SA(t)$, for varying values of t and τ, the time the decisionmaking process has been in progress and the experience of the team in terms of the number of months it has worked as a team, respectively.

Note that teams that have been operating together less than approximately two weeks have very little situational awareness, regardless

RAND *MR1467-C.4*

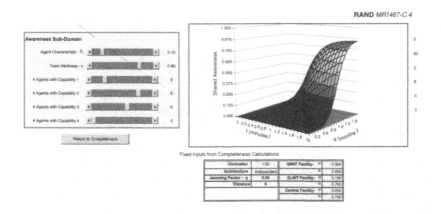

Figure C.4—Infoview Cognitive Domain GUI

of how long the operation extends. Also note that shared situational awareness increases continually in both directions, as the team gains training and experience from the ongoing operation. However, the rate of increase tapers dramatically at around 6 weeks of training and about 1 minute of operational experience. This is primarily due to the behavior of information quality.

ANALYSIS

The spreadsheet model can be used to conduct analysis. A basic assumption for this version of the model is that the mathematical representations for completeness in all domains accurately reflect experience with the components of each domain and that the mathematical representations for transactive memory and individual situational awareness adequately reflect their effects on shared situational awareness. These assumptions remain to be tested.

Typically, a research question will focus on the effects of one or more of the model parameters on information quality and, for a fixed decisionmaking team, the effect on shared situational awareness. For example, we may wish to examine the effects of jamming or the level of automation and sensor control on completeness given a decisionmaking team composed as in the example above. The other parameters may remain fixed at some appropriate level or may take on a range of settings. All this raises the issue of combinatorial analysis. Since each of the slider bars in the model is continuous, an infinite number of combinations are possible. Consequently, some bounds must be placed on the selections.

Suppose the research question were: "How does jamming affect completeness and what is its subsequent effect on shared situational awareness?" We would further suppose the following: (1) that two sensor architectures are of interest: mixed and independent; (2) that there is very little automation and sensor control in IMINT fusion ($\sigma = 0.2$ and $\alpha = 0.5$); (3) that there is considerable automation but little control in ELINT fusion ($\sigma = 0.2$ and $\alpha = 0.95$); (4) that there is considerable automation and control at the central processing facility ($\sigma = 1$ and $\alpha = 0.95$); (5) that there is no "onboard" fusion at the sensors (for all facilities); and (6) there are no occlusions ($k = 1$ for all sensors).

Since we are examining the effects of jamming, we must also specify the range of the sensors to the targets. Suppose we are content to set three ranges: near (0.2 km), medium (1.2 km), and distant (2 km). Since the fusion facility parameters are fixed, as is the level of occlusions, the total number of cases to be examined is six (two architectures and three ranges). Figures C.5 and C.6 summarize the results.

Now let us extend the analysis to the cognitive domain. Clearly, for each of the six cases, several additional cases might be postulated simply by assuming different team hardness parameters and threshold values and by varying the competence of the team members. For simplicity, however, we will hold these fixed to the levels discussed above. Examining only the independent architecture cases depicted in Figure C.6 yields the three shared situational awareness depictions in Figure C.7.

RESULTS

It is clear that, regardless of the level of jamming, the completeness is never more than about 55 percent for the best case and is only slightly better for the independent architecture cases. As expected, the farther the sensor suite is from the target, the lower the system completeness. In all cases, as jamming increases, completeness falls off rapidly.

Note also that shared situational awareness in the independent architecture case is sensitive to the sensor distances. However, for

RAND *MR1467-C.5*

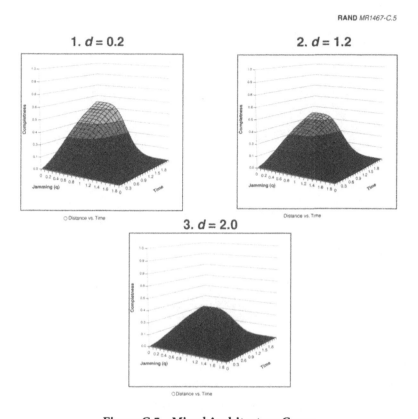

Figure C.5—Mixed Architecture Cases

short distances (0.2 and 1.2 km), the differences are minimal (approximately 0.8 and 0.76 at peak performance for the two cases, respectively). However, as the distance extends to 2 km, shared situational awareness drops off dramatically (to approximately 0.13), suggesting that additional ranges between 1.2 and 2.0 km should be investigated.

A result such as this would suggest that the research question was wrong. Although jamming does have an effect on system completeness, it appears that the levels of automation and sensor control at the fusion facilities are much more crucial. Consequently, it might have been more productive to ask how the level of automation and sensor control at the fusion facilities affects information completeness and shared situational awareness.

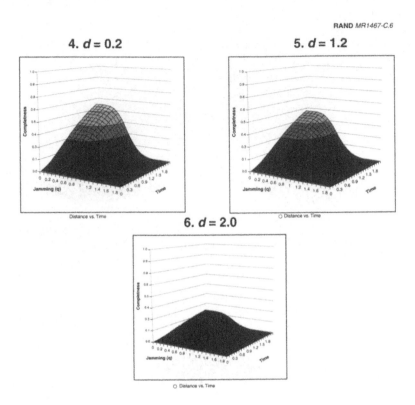

Figure C.6—Independent Architecture Cases

RAND *MR1467-C.7*

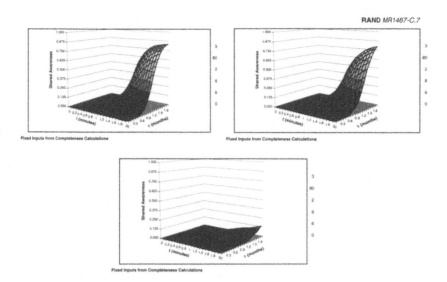

Figure C.7—Independent Architecture Shared Situational Awareness

BIBLIOGRAPHY

Alberts, D., et al., *Understanding Information Age Warfare*, Washington, D.C.: DoD C4ISR Cooperative Research Program, 2001.

Alberts, D. S., J. J. Garstka, and F. P. Stein, *Network Centric Warfare: Developing and Leveraging Information Superiority*, 2nd ed., Washington, D.C.: DoD C4ISR Cooperative Research Program, 2000.

Allard, C. K., *Command, Control, and the Common Defense*, New Haven, Conn.: Yale University Press, 1990.

Ash, R. B., *Information Theory*, Mineola, N.Y.: Dover Publications, 1990.

Askins, S. W., *The Black Box Conundrum of Network Centric Warfare*, thesis, Newport, R.I.: Naval War College, 2000.

Ayyub, B., and R. McCuen, *Probability, Statistics and Reliability for Engineers*, Boca Raton, Fla.: CRC Press, 1997.

Bandura, A., *Self-Efficacy: The Exercise of Control*, New York: W. H. Freeman and Company, 1997.

Bankes, S., "Exploratory Modeling for Policy Analysis," Santa Monica, Calif.: RAND Corporation, RP-211, 1993. (Reprinted from *Operations Research*, Vol. 41, No. 3, May–June 1993.)

Blackman, S., *Multiple-Target Tracking with Radar Applications*, Norwood, Mass.: Artech House, 1986.

Blahut, R. E., *Principles and Practice of Information Theory*, Boston: Addison-Wesley, 1988.

Blake, I. F., *An Introduction to Applied Probability*, Melbourne, Fla.: Krieger Publishing Company, 1987.

Bowerman, B., and R. O'Connell, *Linear Statistical Models: An Applied Approach*, 2nd ed., Boston: PWS-Kent Publishing Co., 1990.

Brooks, A., S. Bankes, and B. Bennett, *Weapon Mix and Exploratory Analysis: A Case Study*, Santa Monica, Calif.: RAND Corporation, DB-216/2-AF, 1997.

Brooks III, F. P., *The Mythical Man Month: Essays on Software Engineering, Anniversary Edition*, New York: Addison-Wesley, 1995.

Brown, S. M., E. Santos, Jr., S. Banks, and M. Oxley, "Using Explicit Requirements and Metrics for Interface Agent User Model Correction," *Proceedings of the Second International Conference on Autonomous Agents*, Minneapolis, Minn., May 10–13, 1998, pp. 1–7.

Builder, C. H., S. Bankes, and R. Nordin, *Command Concepts: A Theory Derived from the Practice of Command and Control*, Santa Monica, Calif.: RAND Corporation, MR-775-OSD, 1999.

Campbell, N., *What Is Science*, Dover Publications, 1921. Excerpts reprinted with commentary in Neuman (1988), pp. 1,769–1,801.

Cebrowski, A. K., "Network-Centric Warfare: Its Origin and Future," *U.S. Naval Institute Proceedings*, Vol. 124, No. 1, January 1998, pp. 28–35.

Clark, H. H., *Using Language*, New York: Cambridge University Press, 1996.

Clark, H. H., and S. E. Brennan, "Grounding in Communication," in L. B. Resnick, J. M. Levine, and S. D. Teasley, eds., *Perspectives on Socially Shared Cognition*, Washington, D.C.: American Psychological Association, 1991, pp. 3,127–3,149.

Coakley, T. P., ed., *Issues of Command and Control*, Washington, D.C.: National Defense University, 1991.

"The Cooperative Engagement Capability," *Johns Hopkins APL Technical Digest*, Vol. 16, No. 4, 1995.

Darilek, R., W. Perry, J. Bracken, J. Gordon, and B. Nichiporuk, *Measures of Effectiveness for the Information-Age Army*, Santa Monica, Calif.: RAND Corporation, MR-1155-A, 2001.

Dhillon, B. S., *Mechanical Reliability: Theory, Models and Applications*, AIAA Education Series, Reston, Va.: American Institute of Aeronautics and Astronautics, 1988.

Duda, R. O., and P. E. Hart, *Pattern Classification and Scene Analysis*, Hoboken, N.J.: John Wiley, 1973.

Endsley, M. R., "Situation Awareness Global Assessment Techniques (SAGAT)," *Proceedings of the IEEE National Aerospace Conference*, Vol. 3, Hawthorne, Calif., May 23–27, 1988, pp. 789–795.

Friedell, M. F., "On the Structure of Shared Awareness," University of Michigan, Department of Sociology, Center for Research on Social Organization, working paper no. 27, April 1967.

Gilder, G., "Metcalf's Law and Legacy," *Forbes ASAP*, September 13, 1993.

Gordon, A. D., *Classification*, 2nd ed., New York: Chapman and Hall, 1999.

Greenhalgh, C., S. Benford, and G. Reynard, "A QoS Architecture for Collaborative Virtual Environments," *Proceedings of Seventh ACM International Conference on Multimedia*, Orlando, Fla., October 30–November 5, 1999, pp. 121–130.

Hall, R. W., A. Mathur, F. Jahanian, A. Prakash, and C. Rassmussen, "Corona: A Communication Service for Scalable, Reliable Group Collaboration Systems," *Proceedings of the ACM 1996 Conference on Computer Supported Cooperative Work*, Boston, November 16–20, 1996, pp. 140–149.

Halmos, P., *Measure Theory*, Princeton, N.J.: Van Nostrand, 1950.

Honderich, T., ed., *The Oxford Companion to Philosophy*, New York: Oxford University Press, 1995.

Institute of Electrical and Electronics Engineers, *IEEE Standard Dictionary of Terms*, 6th ed., New York: IEEE Publications, June 2000.

Jackson, B. W., and D. Thoro, *Applied Combinatorics with Problem Solving*, Boston: Addison-Wesley, 1990.

Keithley, H., *Multi-INT Fusion Performance*, Arlington, Va.: Joint C^{4}ISR Decision Support Center, DSC-00-02, April 2000.

Kraut, R. E., M. Miller, and J. Siegel, "Collaboration in Performance of Physical Tasks: Effects on Outcomes and Communication," *Proceedings of the ACM 1996 Conference on Computer Supported Cooperative Work*, Boston, November 16–20, 1996, pp. 57–66.

Kreyzig, E., *Introductory Functional Analysis with Applications*, New York: Wiley, 1978.

Lewis, D. K., "Scorekeeping in a Language Game," *Journal of Philosophical Logic*, Vol. 8, 1979, pp. 339–359.

Mandviwalla, M., and L. Olfman, "What Do Groups Need? A Proposed Set of Generic Groupware Requirements," *ACM Transactions on Computer-Human Interaction*, Vol. 1, No. 3, 1994, pp. 245–268.

McDaniel, S. E., "Providing Awareness Information to Support Transitions in Remote Computer-Mediated Collaboration," *Conference Proceedings on Human Factors in Computing Systems*, Amsterdam, The Netherlands, April 24–29, 1993, pp. 269–276.

Mitchell, R., "Naval Fire Support: Ring of Fire," *U.S. Naval Institute Proceedings*, Vol. 123, No. 11, November 1998, p. 54.

Mitchie, J. K., ed., *Multisensor Fusion for Computer Vision, Proceedings of the NATO Advanced Research Workshop on Multisensor Fusion for Computer Vision*, Grenoble, France, June 26–30, 1989, New York: Springer-Verlag, 1991.

Nardulli, B. R., W. L. Perry, B. Pirnie, J. Gordon, and J. G. McGinn, *Disjointed War: Military Operations in Kosovo, 1999*, Santa Monica, Calif.: RAND Corporation, MR-1406-A, 2002.

Neter, J., and W. Wasserman, *Applied Linear Statistical Models*, Chicago: R. D. Irwin, 1974.

Neuman, J. R., ed., *The World of Mathematics: A Small Library of the Literature of Mathematics from A'h-mosé the Scribe to Albert Einstein*, Vol. 3, Tempus Books, 1988.

Nichols, P., *"U.S. 2025 Information Dominance,"* briefing, Fort Monmouth, N.J.: U.S. Army Communications-Electronics Command Research Development and Engineering Center, June 1999.

Office of the Secretary of Defense, Information Superiority Working Group, *Understanding Information Warfare*, Washington, D.C., 2000.

Patel, U., M. J. D. Cruz, and C. Holtham, "Collaborative Design for Virtual Team Collaboration: A Case Study for Jostling on the Web," *Proceedings of the Conference on Designing Interactive Systems: Processes, Practices, Methods, and Techniques*, Amsterdam, The Netherlands, August 18–20, 1997, pp. 289–300.

Pecht, M., ed., *Product Reliability, Maintainability and Supportability Handbook*, Boca Raton, Fla.: CRC Press, 1995.

Perry, W., "Knowledge and Combat Outcomes," *Military Operations Research Journal*, Vol. 4, No. 1, 2000.

Perry, W., and J. Moffat, "The Value of Information on the Outcome of Maritime Operations: The Use of Experts in Analysis," *Proceedings of 11th International Symposium on Operational Research*, Shrivenham, Wiltshire, United Kingdom, September 5–9, 1994.

Perry, W. and J. Moffat, "Measuring Consensus in Decision Making: An Application to Maritime Command and Control," *Journal of the Operational Research Society*, Vol. 48, No. 4, 1997, pp. 383–390.

Perry, W., and J. Moffat, "Developing Models of Decision Making," *Proceedings of 12th International Symposium on Operational Research*, Shrivenham, Wiltshire, United Kingdom, September 5–8, 1995, and *Journal of the Operational Research Society*, Vol. 48, No. 5, 1997, pp. 457–470.

Perry, W., and J. Moffat, "Measuring the Effect of Knowledge in Military Campaigns," *Journal of the Operational Research Society*, Vol. 48, No. 10, 1997, pp. 965–972.

Perry, W., and T. Sullivan, *Modeling Information Processing with Rec-ommendations for JWARS*, unpublished draft, Santa Monica, Calif.: RAND Corporation, October 1999.

Pryor, J. B., and T. M. Ostrom, "Social Cognition Theory of Group Processes," in B. Mullen and G. R. Goethals, eds., *Theories of Group Behavior*, New York: Springer-Verlag, 1987, pp. 147–183.

Reber, A. S., *The Penguin Dictionary of Psychology*, New York: Penguin Books, 1995.

Russell, D. M., M. J. Stefik, P. Pirolli, and S. K. Card, "The Cost Structure of Sensemaking," *Conference Proceedings on Human Factors in Computing Systems*, Vancouver, Canada, April 13–18, 1996, pp. 57–58.

Signori, D., et al., "A Conceptual Framework for Network Centric Warfare," presentation to the Technical Cooperation Program's Network Centric Warfare/Network Enabled Capabilities Work-shop, Arlington, Va., December 17, 2002. Online at http://www.dodccrp.org (as of September 26, 2003).

Smith, D. R., *The Effect of Transactive Memory and Collective Efficacy on Aircrew Performance*, dissertation, University of Washington, 1999, Ann Arbor, Mich.: University Microfilms, No. 05858900 388261,1999.

Stalnaker, R. C., "Assertion," in P. Cole, ed., *Syntax and Semantics 9: Pragmatics*, New York: Academic Press, 1978, pp. 315–332.

Stark, H., and J. W. Woods, *Probability, Random Processes and Esti-mation Theory for Engineers*, Englewood Cliffs, N.J.: Prentice-Hall, 1986.

Stein, F. P., "Observations on the Emergence of Network Centric Warfare," Office of the Assistant Secretary of Defense (NII), Com-mand and Control Research Program, June 1998. Online at http://www.dodccrp.org/steinncw.htm (as of September 22, 2003).

Taha, H., *Operations Research: An Introduction*, 3rd ed., New York: Macmillan, 1976

Ulhoi, J. P., and U. E. Gattiker, "The Nature of Technological Paradigms: A Conceptual Framework," in R. C. Dorf, ed., *The Technology Management Handbook,* New York: CRC Press/IEEE Press, 1999, pp. 7-87 through 7-93.

U.S. Army Digitization Office, *Providing the Means,* Washington, D.C.: Office of the Chief of Staff Army, 1994.

U.S. Department of Defense, *Department of Defense Dictionary of Military and Associated Terms,* Washington, D.C., Joint Publication 1-02, September 5, 2003.

Vin, H. M., V. Rangan, and S. Ramanathan, "Hierarchical Conferencing Architecture for Inter-Group Multimedia Collaboration," *Conference Proceedings on Organizational Computing Systems,* Atlanta, Ga., November 5–8, 1991, pp. 43–54.

Waltz, E., and J. Llinas, *Multisensor Data Fusion,* Norwood, Mass.: Artech House, 1990.

Wegner, D. M., "Transactive Memory: A Contemporary Analysis of the Group Mind," in B. Mullen and G. R. Goethals, eds., *Theories of Group Behavior,* New York: Springer-Verlag, 1987, pp. 185–208.

Wellens, A. R., *Assessing Multi-Person and Person-Machine Distributed Decision Making Using and Extended Psychological Distancing Model,* Wright-Patterson AFB, Ohio: Harry G. Armstrong Aerospace Medical Research Laboratory, Human Systems Division, Air Force Systems Command, AAMRL-TR-90-006, February 1990.